how
to
know
the

spiders

The **Pictured Key Nature Series** has been published since 1944 by the Wm. C. Brown Company. The series was initiated in 1937 by the late Dr. H. E. Jaques, Professor Emeritus of Biology at Iowa Wesleyan University. Dr. Jaques' dedication to the interest of nature lovers in every walk of life has resulted in the prominent place this series fills for all who wonder **"How to Know."**

John F. Bamrick and Edward T. Cawley
Consulting Editors

The Pictured Key Nature Series

How to Know the
 AQUATIC INSECTS, Lehmkuhl
 AQUATIC PLANTS, Prescott
 BEETLES, Arnett-Jaques, Second Edition
 BUTTERFLIES, Ehrlich
 CACTI, Dawson
 EASTERN LAND SNAILS, Burch
 ECONOMIC PLANTS, Jaques, Second Edition
 FALL FLOWERS, Cuthbert
 FERNS, Mickel
 FRESHWATER ALGAE, Prescott, Third Edition
 FRESHWATER FISHES, Eddy-Underhill, Third
  Edition
 GILLED MUSHROOMS, Smith-Smith
 GRASSES, Pohl, Third Edition
 IMMATURE INSECTS, Chu
 INSECTS, Bland-Jaques, Third Edition
 LAND BIRDS, Jaques
 LICHENS, Hale, Second Edition
 LIVING THINGS, Jacques, Second Edition
 MAMMALS, Booth, Third Edition
 MARINE ISOPOD CRUSTACEANS, Schultz
 MITES AND TICKS, McDaniel
 MOSSES AND LIVERWORTS, Conard-Redfearn,
  Third Edition
 NON-GILLED FLESHY FUNGI, Smith-Smith
 PLANT FAMILIES, Jaques
 POLLEN AND SPORES, Kapp
 PROTOZOA, Jahn, Bovee, Jahn, Third Edition

 ROCKS AND MINERALS, Helfer
 SEAWEEDS, Abbott-Dawson, Second Edition
 SEED PLANTS, Cronquist
 SPIDERS, Kaston, Third Edition
 SPRING FLOWERS, Cuthbert, Second Edition
 TREMATODES, Schell
 TREES, Miller-Jaques, Third Edition
 TRUE BUGS, Slater-Baranowski
 WATER BIRDS, Jaques-Ollivier
 WEEDS, Wilkinson-Jaques, Third Edition
 WESTERN TREES, Baerg, Second Edition

how
to
know
the
spiders

Third Edition

B.J. Kaston
San Diego State University

The Pictured Key Nature Series
Wm. C. Brown Company Publishers
Dubuque, Iowa

Contents

Preface to the Third Edition

During the past few years, in the United States and Canada, there has been a tremendous surge of interest in the study of Arachnids. This is evidenced in part by the organization of an American Arachnological Society, with its Journal of Arachnology. But more especially, there has been an increase in the number of publications containing taxonomic revisions of different genera, and even of families. Our knowledge of these groups has been greatly expanded, and the ranges of many heretofore little known species have been greatly extended. This applies particularly to the western half of the continent with its great variety and numbers of species. Again I have had to make some name changes because of synonymy, have corrected some errors, and here and there have supplied new data. With the inclusion of 13 additional genera and of 121 additional species this book now covers almost twice as many as there were in the original edition. At the suggestion of some users of previous editions I have added a key to all the Orders of Arachnida, which enables one to see how the spiders fit in with their non-aranean relatives.

I am indebted to the following individuals for the loan of specimens and help of various kinds: Dr. W.J. Gertsch, Dr. H.W. Levi, Dr. Allen R. Brady, Dr. H.K. Wallace, Dr. Norman I. Platnick, David B. Richman, Saul Frommer, Leonard S. Vincent, Steven C. Johnson, and Wendell Icenogle. The additional drawings were prepared by Oscar Padilla.

San Diego, California   B.J. Kaston

Introducing Spiders

What is most often associated with a spider in the minds of the majority of people? Its web! The very name, *spider,* refers to its spinning habit, though the making of a snare is not the only use to which the spider puts its silk. The wandering spiders build no snares yet use the silk in various ways. Most spiders as they move about pay out a *"drag-line"* behind them, and it is this type of line which supports the spider should it lose its footing or jump from a support. At frequent intervals the drag-line is fastened to the substratum by a large number of looped threads. This is called an *"attachment disc"* and it is from this point that the spider is supported should it drop on its drag-line.

Prey may be wrapped by a *"swathing band"* composed of numerous strands. Many spiders build a nest, or retreat, which may be a tube open at both ends, a silk-lined excavation, or an inverted cup near the web. The same or similar nests may be constructed for molting, mating, or hibernating. The females make *egg sacs,* and the males *sperm webs.* The snares may be *irregular meshes,* as in the Theridiidae, *sheet webs,* as in the Linyphiidae, *funnel webs,* as in the Agelenidae, *orb webs* as in Araneidae, Theridiosomatidae, and Tetragnathidae, or else combinations of these.

One of the most interesting uses to which silk is put by some spiders is a "balloon." The spider climbs up high on a blade of grass, on a fence, or pole, and facing the wind stands on the tips of the tarsi and tilts the abdomen upwards. From the spinnerets are emitted threads which are paid out as the air currents pull, until the buoyancy of the parachute is enough to support the spider, which releases its hold and is carried away in the breeze (fig. 1).

Figure 1. Spider getting ready to balloon.

The eggs are always laid within a cocoon of silk, though in some like *Pholcus* and *Scytodes,* both of which are house spiders, this may be quite scanty (fig. 2). The eggs of running spiders are laid upon a sheet of silk which is then wrapped around the egg mass, or another sheet is made over the eggs. Some spiders enclose a padding of silk before closing up the egg sac. In some the outer covering is thin and meshy, in others quite tough, and in still others papery in texture. It may be fastened to the surface of bark, on a twig, under a leaf or stone; or it may be suspended in the web or retreat, either with a stalk or without; or it may be carried around by the mother attached to her spinnerets (fig. 3), held by her chelicerae or legs, or held under her sternum (fig. 4). The shape may be spherical, as is usual, or commonly like a plano-convex or biconvex lens, or less commonly pear shaped (fig. 5).

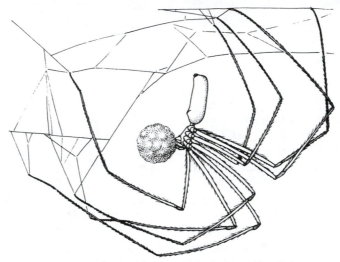

Figure 2. *Pholcus phalangioides,* in web, holding her ball of eggs.

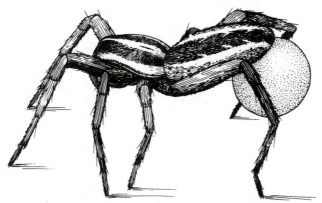

Figure 3. *Lycosa* with egg sac attached to spinnerets.

Figure 4. *Pisaurina* with egg sac held under her cephalothorax.

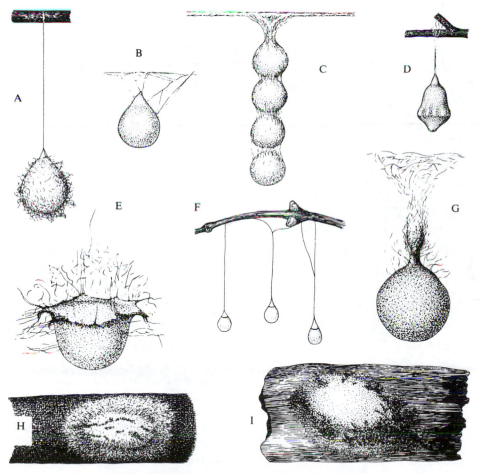

Figure 5. Some spider egg cocoons. A, *Ero furcata;* B, *Argiope argentata;* C, *Mecynogea lemniscata;* D, *Conopistha rufa;* E, *Argiope trifasciata;* F, *Theridiosoma radiosa;* G, *Argiope aurantia;* H, *Tetragnatha elongata;* I, *Agelenopsis* sp.

Many species leave their egg sacs and never see their offspring. On the other hand, among the thomisids, salticids, many clubionids, and gnaphosids the mother mounts guard over the sac until the offspring emerge. The lycosid carries her young about on her body and the pisaurid fashions a nursery web and mounts guard outside, a little below the nursery, until sometime after the spiderlings have emerged (fig. 6).

The number of eggs in a sac varies widely; in a few cases only one or two, in some up to 25 or 30, in many from 100 to 300, and in one instance 2652 were counted in a sac made by the orb weaver, *Araneus trifolium*. Some spiders make more than one sac (up to 20 or more) and occasionally three or four may be seen in a web at the same time.

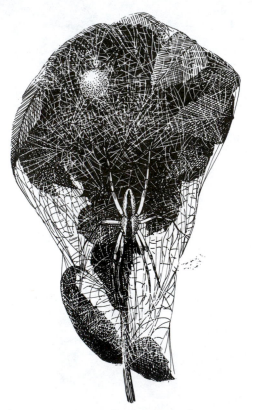

Figure 6. *Pisaurina* standing guard below her nursery web.

The number of times a spiderling molts to attain maturity varies with the species. In general, the smaller species molt fewer times and the larger more, with males molting fewer times than the females.

Like many other Arthropoda spiders are able to lose one or more appendages without being greatly inconvenienced. If a spider is grasped by a leg, the appendage may be *autotomized* and the spider escape. Provided another molt is not too near the missing appendage can be regenerated, though it will be somewhat smaller than the normal.

THEIR FOOD

Spiders are exclusively carnivorous, and in fact generally seize only live animals, as the spider's attention is attracted by the movement of the prey. Very occasionally one learns of a snake or mouse accidentally caught in the snare of a spider, but ordinarily the spiders of our region feed on insects. Certain pisaurids, however, our *Dolomedes* for example, sometimes capture small fish, such as minnows, and the tarantulas and trap-door spiders have been known to sometimes catch frogs, lizards, snakes, mice and birds. Most species are not particular as to the insects eaten but will take whatever happens to come their way. However, it is true that some species will not accept everything. Only a few eat ants, many refuse wasps and hornets, and many cast bugs and beetles out of their webs. Some spiders have been known to eat bees, but this undesirable behavior is counterbalanced by the fact that they also eat enormous numbers of harmful insects such as mosquitoes.

In the case of the trap-door and wandering spiders, the prey is not wrapped in silk, but simply seized, held by the mouthparts, and eaten. Theridiids and orb weavers generally enswathe the insect caught in their snares, then bite it, and later drag it to the hub, or to the retreat, to be eaten (fig. 7). Members of the genus *Scytodes* secure their prey with a mucilaginous material made in the poison glands. This is ejected forcibly and hardens almost immediately, thus fastening the insect to the substratum.

There are two ways in which spiders ingest food. Those with weak jaws puncture the body of the insect with their fangs and then slowly alternate between injecting digestive fluid through this hole and sucking back the liquified tissues, until there remains but an empty shell. The tarantulas, wolf spiders, large orb weavers, and others with strong jaws mash the insect to a pulp between the jaws, as the digestive fluid is regurgitated over it. Only a small mass of undigestible material, such as the chitinous elements, will remain to be discarded. To finicky humans this process may seem repulsive, but to anyone who has seen a preying mantid voraciously devour a grasshopper, the spider's method will seem by contrast positively dainty.

The spider's appetite may often appear insatiable, the abdomen swelling to accommodate the added food. Most species can survive long periods of fasting, and although many can do without water for weeks some species will die if deprived of moisture for only a few days.

Spiders may feed on other spiders, and because of this tendency to cannibalism a social or communal life is hardly to be expected. Yet instances of commensalism are known. The small spiders of the genus *Conopistha* usually inhabit the snares of larger spiders and consume small insects which are neglected by the web's owner. In the case of hibernating, a number of salticid, thomisid, and clubionid individuals of the same species often come together in large groups at the advent of cold weather and spend the winter in a pseudo-social "flock."

COURTSHIP

Among the remarkable phenomena occurring in spiders ranks the peculiar behavior associated with mating. When the male matures considerable effort is put forth in locating the female; which means in the case of snare-building species that it now becomes a wanderer like the hunting spiders. When the female is approached certain preliminaries are usually

Figure 7. Spider wrapping prey.

engaged in before mating takes place. These courtship maneuvers are usually started by the male and continued by him, though in some cases the female may also take part after she has reached a certain pitch of excitement. As a rule, in the web-building species the male signals by tweaking the threads of the female's snare. In addition there may be movements of palpi and abdomen, a sort of dance that varies slightly with the species. It is in the wandering spiders that courtship is most marked. This is particularly true of those families with relatively keen eyesight. Here the males may dance before the females, wave their palpi, or legs, or both, and strike peculiar attitudes (fig. 8). It has been assumed that all of this serves to gain recognition by the female, as well as to stimulate her. The male's behavior eventually leads to a lulling of the female's normal instinct to consider him as a possible morsel to be eaten

and she submits to his advances, subdued by the ardour of the sexual impulse. It is interesting to note that the male of a European pisaurid actually presents the female with insect food, thus ensuring his own safety.

It is a popular misconception that the male is always killed by the female after the mating act is over. Actually, in only a very few species can this be considered the general rule. For the great majority the sexes separate peaceably and the male may even mate again later with the same or other females. Numerous instances are known of male and female sharing the same web, or same retreat, for considerable periods of time. It is to be expected that the male will die sooner than the female. Though the females of some species will live for two or three, or even more, years, many die shortly after laying their eggs and do not survive the winter.

Figure 8. Spiders courting.

Where to Find Them

They may be found almost everywhere; on or near water, in or on the ground, from underground caves to the tops of mountains. In fact salticid spiders have been taken as high as 22,000 feet on Mt. Everest, the highest elevation at which any animal has ever been taken. Ballooning spiders have actually been collected from airplanes at an elevation of 5,000 feet. Some kinds of spiders live inside human habitations, others frequent the outside of structures. Tall and low plants have spider tenants, as do the dead leaves on forest floors and the curled, dried leaves on trees in winter. Under bark, under stones, under fallen logs—these are only a few of their varied habitats. There may be many different kinds even in an area of limited size; for example over 650 species are known from Connecticut, a very small State. The number of individuals may be prodigious, as one worker found a concentration of 407,000 per acre of clay meadow, and another over 2,200,000 per acre in a grassy field.

Of the ground spiders, some, like *Geolycosa,* dig holes in which they remain for the duration of life, except for the brief period when the male ventures out to seek a mate. The trap-door spiders of the southern and western states also dig holes, but their deluxe abodes are lined with silk, and possess a silk-hinged door that fits snugly over the entrance. The silk-lined tunnel of *Atypus* extends partly into the ground, and partly along the surface or against a tree. The wolf spiders and tarantulas may make use of shallow holes in which to hide. Many of the gnaphosids, and some clubionids run about over the ground and have been found under stones in meadows and in woods. A large number of spiders are found among and under loose rock formations.

Many spiders prefer dark and shaded locations, where the humidity is high. They may therefore on that account be found in caves or deserted mines, and also in the cellars of buildings. In amongst the dead leaves and litter on forest floors occur many species, which, being usually dark in color, and blending with the background, are often difficult to see when they are not moving.

Most *Pirata* and *Dolomedes* species are found along the edges of streams and ponds, over the surface of which they run, and beneath the surface of which they can dive. *Pachygnatha* and *Tetragnatha* species also prefer water courses, but are usually found on the shrubbery which overhangs the ponds or streams.

Other species are found in tall grass, on bushes and trees. Some run over the branches and trunk, and hide under loose bark and in crevices. Snares may be built among twigs, and many linyphiids, theridiids, and orb weavers construct their webs in tall grass, bushes and tree foliage. *Tibellus* and *Thanatus* run along grass stems, clubionids, anyphaenids, and salticide hunt from leaf to leaf, and *Misumena* lives among flowers, waiting in ambush for insects to come within reach. These "flower" spiders are often colored like their backgrounds and to a certain extent can change their color from white to yellow or back again.

Many spiders are not only colored like their surroundings but are of a peculiar body form resembling inanimate objects. Some, like *Cyclosa,* resemble a piece of bird dung; others, like *Hyptiotes,* look like the buds of the plants on which the web is built; and many with elongated bodies and legs resemble pieces of straw and grass. Thus even though there may be a goodly number of spiders in an area some species may be seldom encountered by humans because of this camouflage, or because of their habit of remaining hidden in the ground.

Some spiders resemble other animals and of these mimics the ant-like species are most common. Examples are known in several families and often the mimicry extends not only to the body form but also to the behavior, the spiders moving about with anterior legs elevated like a pair of antennae.

How to Collect
and Preserve Them

Having a knowledge of the habits of spiders, one is prepared to look for them in a variety of habitats. In building a collection one must look in as many kinds of places as possible and at all times of the year. One of the easiest ways to get large numbers is by using a sweeping net[1] through tall grass and weeds and picking out the spiders from among the insects, leaves and debris that will be gathered with them. Bushes should be beaten and small trees may be shaken, after first putting down a light-colored cloth to make visible the specimens that drop. In walking through the woods webs should be looked for, especially when moving toward the sun, and often ground dwelling species will appear on the path. Webs are especially easy to see in the morning when laden with dew. Look under logs and stones in wooded areas and pastures. One should also watch for the holes of burrowing forms in sod and in the sand at the seashore.

It requires sharp eyes to locate the purse webs, made by the Atypidae along the lower trunks of trees, and the domiciles of the trap-door spiders in shaded ravines or steep canyon walls, or even the desert floor. Often spiders may be hiding in driftwood and litter along the beach, on flowers, in the cracks of fences (stone or board) and in old unused outbuildings. For spiders that live in forest floor humus, or in moss, the material may be placed in a coarse sieve (a French frying basket makes an excellent one for this purpose) and the spiders shaken down on to a piece of cloth. A new white cloth is too dazzling in the sun, therefore it is best to use one that is dirty, or else a light gray. Instead of using a sieve, one may place the mass of dead leaves and litter in the center of a large canvas sheet and slowly remove it bit by bit, watching carefully for any spiders that attempt to escape by scurrying toward the edge of the sheet.

One technique for collecting ground forms, especially males of wandering species, is to set out "pit-fall" traps (fig. 9). These can be large cylindrical open cans, or white plastic buckets, set into the ground so that the top is flush with the ground level. The can should be loosely covered with a large flat stone, or a piece of plywood supported by low small stones, to keep out large predators and rain. To keep specimens from eating each other, and to prevent decomposition of those that die, it is best to pour into the trap some ethylene glycol (which does not evaporate readily) to a depth of about one or two inches. This will kill and temporarily preserve specimens that fall in. Of course, if one wishes to keep the specimens alive, as for the rearing of juveniles, the ethylene glycol must be omitted, and the traps will have to be checked more frequently.

1. See the description in Bland-Jaques "How to Know the Insects."

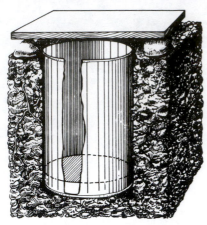

Figure 9. Pit trap for collecting spiders that wander over the ground.

During the winter when many spiders are torpid in the forest litter, cloth or plastic sacks filled with this material can be brought into the laboratory and sifted at leisure over a table after the spiders have had an opportunity to warm up and become active.

Specimens may be picked up with forceps, or the fingers, and for tiny species a camel's hair brush moistened with the preserving fluid is satisfactory. These may then be transferred to the collecting jar of preserving fluid. It is also convenient to collect in a dry shell vial of the proper size. This method has the advantage that it is least apt to injure the specimen, and also allows the collector to easily examine the specimen before killing it, or to conveniently transport it back to the laboratory alive if desired.

Since web spiders usually display a tendency to drop when disturbed, the open vial should be held *below* ready to receive the specimen on its way down. The vial is placed over ground specimens and a leaf or piece of paper passed between the mouth of the vial and the ground thus barring escape while the vial is raised. An assortment of different sized vials should be carried. They do not need to be very large, for the spiders draw up their legs and thus may be induced to enter a vial of surpris-ingly small diameter. Vials up to an inch in diameter will do for all but tarantulas, trap-door spiders, the largest *Lycosa* and *Dolomedes*. For these latter a small tumbler will do.

Often difficulty may be experienced in attempting to shake the spider from a dry vial into the alcohol. Spiders are repelled by alcohol fumes and stubbornly resist being dislodged from the dry vial. In that case, since they are hardly at all repelled by cyanide it becomes a simple matter to first quiet them by placing them for a few minutes in an ordinary entomological killing tube,[2] after which they may be transferred to alcohol. The cyanide tube can also be used conveniently in picking out specimens from the sweeping net. Since they begin to shrivel quickly, dead specimens should be placed in preserving fluid as soon as possible.

No survey of an area is complete without some night collecting being done. Many orb weavers which hide during the day can then be seen at the hub, or repairing their webs. Many clubionids can be seen making their way over the foliage, but perhaps the most outstanding success in night collecting is with the Lycosidae. One wears on the forehead an electric torch, like a miner's lamp, tilted so as to throw a spot of light about 12 to 20 feet away (fig. 10). The eyes of the wolf spiders reflect a greenish light, which, after a little experience, becomes easily recognizable. One soon learns how to slowly approach the specimen so as to keep its eyes always in sight; and since most spiders remain motionless during this time they are easily collected. Besides the wolf spiders, many crab spiders and pisaurids can be taken with a light in this way. One of the most spectacular finds is a female lycosid carrying her young. Then one may see the large eyes of the mother, and hundreds of tiny eyes of the young, all shining like jewels on black velvet.

2. See Bland-Jaques "How to Know the Insects."

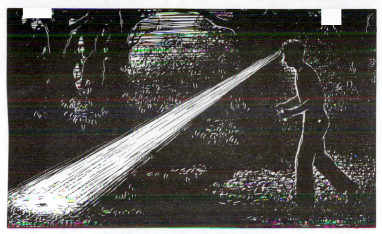

Figure 10. This is well worth trying.

If live specimens are to be kept in the laboratory, as for rearing nearly mature individuals, or for any other purpose, the collector must remember that the cannibalistic habits of spiders necessitates that each be kept in its own container. I have found ordinary drinking tumblers (fig. 11) suitable for all but the very largest species (which may require mason jars). The mouth of the tumbler should be plugged with a wad of cotton that has been wrapped and tied in a piece of cheesecloth. This allows air circulation, prevents the cotton from being pulled apart, and keeps both spider and the introduced insect food from escaping. Similarly

Figure 11. Water tumbler fitted with stopper, used for keeping spiders alive in the laboratory.

the easily obtainable baby-food jars can be used. In recent years I have found that polyurethane foam cut to proper size serves well in place of cotton plugs. While the spider may withstand a long fast, one should add a drop or two of water each day. If one is raising juveniles, which will be molting later on, it is necessary to have in the tumbler some material to which they can cling, as a small twig, a leaf, or even some crumpled tissue paper.

For the permanent preservation of spiders in a collection some workers prefer alcohol undiluted, others use various dilutions down to 70%. There should of course be a large quantity of alcohol for the size of the specimen since the body fluids will dilute the preserving fluid. The colors are not preserved in alcohol, unfortunately, but partial success may be obtained by addition of some formalin. I have found satisfactory a mixture of 10 parts commercial formalin with 90 parts of 60% alcohol. Some workers use mixtures containing acetic acid also. One such mixture contains formalin 12 parts, alcohol 30 parts, glacial acetic acid 2 parts, and distilled water 56 parts. In another mixture there are 61 parts of alcohol, 8 of glacial acetic acid, 5 of glycerine, and 26 of water. It is recommended that specimens be removed from the fixing agent after a day or so and placed in 70% alcohol. Isopropyl will do as well as ethyl, is less expensive, and is more easily obtained.

It is well to place in the jar of preserving fluid with the specimens a paper label bearing the place and date of collection, as well as the collector's name. After the specimens are identified the members of each species are transferred to a smaller vial with a label bearing name and sex.

Parasites and Other Enemies

A number of hymenopterous insects parasitize the egg sacs of, and some are found as *external* parasites on, the bodies of the spiders. There are also Diptera which eat the eggs of spiders, and some which are *internal* parasites in the bodies of the spiders. Many animals which are predacious on insects are predators of spiders as well, and this includes spiders themselves.

Perhaps the most serious enemies are the wasps of the two families, Psammocharidae (Pompilidae) and Sphecidae. The former includes the familiar "tarantula hawk," and the latter includes the mud-daubers. In surveying an area for spiders one should always examine the contents of these mud nests, and remove the spiders "collected" by the wasp for provisioning its nest.

What Are Spiders?

Spiders are not insects. Together with scorpions, whip-scorpions, vinegaroons, pseudoscorpions, solpugids, ticks, mites, and daddy-long-legs they belong to the Class Arachnida. They may be readily separated from the daddy-long-legs (harvestmen) with which they are often confused by the fact that the latter have the abdomen noticeably seg- mented and broadly joined to the cephalothorax, and also lack the spinnerets at the hind end of the abdomen.[1]

Spiders are placed in the Order Aransae, and can be readily distinguished from insects by the following characters:

1. A key enabling separation of the eleven Orders of the Class Arachnida will be found on page 33.

	Spider	Insect
Body regions	Two	Three
Antennae	Lacking	One pair
Legs	Four pairs	Three pairs
Pedipalps	One pair of six segments; modified in male for sperm transfer	Absent
Poison apparatus	Opening on fangs of chelicerae	If present, usually opening at posterior end of abdomen
Wings	Always lacking	Most commonly present
Eyes	Always simple ocelli; most commonly 8, or 6	Commonly compound; sometimes with 2 or 3 ocelli in addition
Silk apparatus	Always present, opening at hind end of abdomen below anus	Only in some larvae and opening on lower lip

	Spider	Insect
Genital pore	On ventral side near anterior end of abdomen	Terminal, just below anus at posterior end of abdomen
Food digestion	Always occurs before swallowing, by regurgitation of enzymes	Usually takes place after swallowing
Development	Direct; no larval stages; spiderlings resemble their parents	May have a metamorphosis with larval and pupal stages, or with nymphs.

As will be seen from Figure 12, the body of a spider is divided into two regions connected by a narrow *pedicel.* The anterior division, or *cephalothorax,* bears the eyes, mouthparts and legs. The posterior division, or *abdomen,* bears the openings of the respiratory, reproductive, and digestive systems, and the external spinning apparatus.

The cephalothorax is covered above by a *carapace,* and below by a *sternum* behind and *labium* in front. While unsegmented the carapace often shows a more or less distinct *cervical groove* marking the boundary between the head portion and the thoracic region. Behind the cervical groove in many species is a conspicuous depression, or a pigmented line called the *dorsal,* or *thoracic, furrow, fovea* or *groove.* Three pairs of less conspicuous *radial furrows* may be present behind the cervical groove.

At the front end of the carapace are the eyes, which are always simple *ocelli* (fig. 13). Most spiders have eight (four pairs) but there are some families with fewer, or rarely the eyes are completely lacking. In most families the eyes are arranged in two rows and the entire space they enclose is the *eye area* or *ocular quadrangle.* The area between the anterior row and the edge of the carapace in front is the *clypeus.* The eyes are generally designated as *anterior medians, anterior laterals, posterior medians,* and *posterior laterals,* and their relative size and spacing are important. The area included by the four medians is called the

median ocular area, and its shape and dimensions are of importance. The eyes may be either all of the same dark or light color *(homogeneous)* or not *(heterogeneous).* The anterior medians are nearly always dark and circular, and the others may be circular, oval, or sometimes triangular, and either clear, or of a pearly lustre, or else dark like the anterior medians. Six-eyed spiders lack these anterior medians.

The rows of eyes are frequently curved. If the curvature is such that the lateral eyes are farther forward than the median eyes the row is said to be *procurved;* if farther back, *recurved.* In the case of the anterior row the spider should be viewed from in front, and the distance from the front edge of the clypeus noted. When the rows are very markedly curved the eyes may then appear to be in three, or even four transverse rows.

Extending below and in front of the clypeus is a pair of *chelicerae.* Each consists of a stout basal section, and a smaller distal segment, the *fang,* articulated at the outer corner. The basal segment may, or may not, have a more or less well developed shiny prominence, or *boss,* at the outer edge near the clypeus. In the tarantulas and their kin the chelicerae extend forward and the fangs are so articulated as to move in a plane more or less parallel to the median plane of the body. In the majority of spiders the chelicerae extend downward, or obliquely forward, with the fangs so articulated as to move in a more or less transverse plane. The groove in which the fang lies when not in

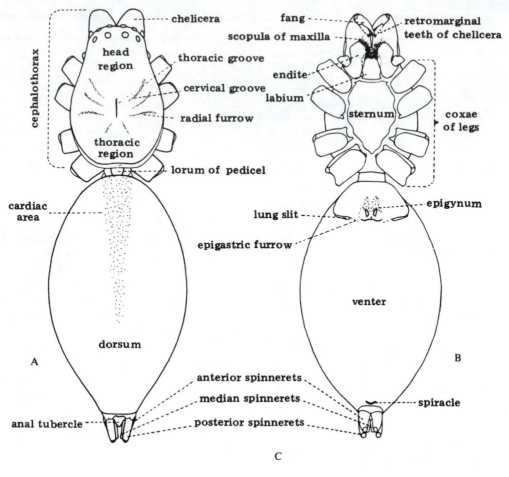

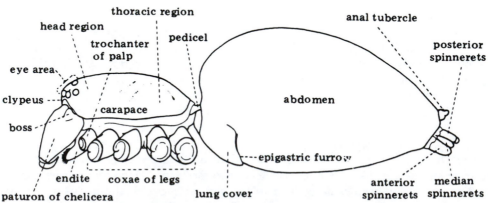

Figure 12. Three views of a spider, without legs, showing parts labeled. A, Dorsal; B, Ventral; C, Lateral.

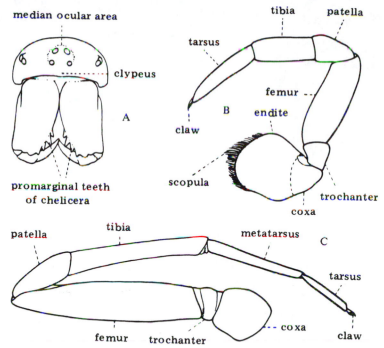

Figure 13. A, Face and chelicerae from in front; B, Pedipalp of female; C, Leg.

use is often armed with teeth, or lobes, of taxonomic importance. Those on the anterior or upper side are the *promarginal teeth,* the others the *retromarginal.* In a few species small, conical, nipple-like tubercles, or *mastidia,* are present on the front of the chelicera. In some groups of spiders each chelicera is provided on its lateral surface with a row of horizontal striae making up the "file" of a stridulating organ, of which the "*pick*" is borne on the pedipalp.

Behind the mouth is a second pair of mouthparts, the *pedipalps.* These are leg-like in appearance but consist of only six, instead of seven segments. The basal segment is expanded in most spiders to form the *maxillae,* or *endites.* The remaining segments constitute the *palp.* Between the two endites is the *labium,* which is usually separate or "free" from the sternum, but in some cases is fused to it so as to be immobile. In certain spiders the distal end of the labium is thickened and strengthened, and is then referred to as *rebordered.*

In females the palpal tarsus is simple and may or may not be armed with a single claw. In mature males, however, the tarsus is modified to carry a more or less complicated *copulatory organ* (figs. 14, 15). Often the tibia, sometimes also the patella (fig. 16), and indeed even the femur, may be provided with *apophyses,* which because of the infinite variety of shapes assumed in different species are of the utmost taxonomic value. In many spiders the tarsus has a bowl-shaped cavity on its ventral surface and hence is called the *cymbium.* In many groups mature males are provided with an appendage, the *paracymbium* (fig. 16), arising from the base of the cymbium. Identification to species and often to genera is based upon the structure of the palpal organ with its many parts, but that is a matter beyond the scope of this book. Because the tarsus is so modified in males it can be used to differentiate mature males from females. In addition males are smaller than females, the abdomen is thinner and the legs longer (compare figures 17A and 17B).

There are four pairs of legs, each of seven

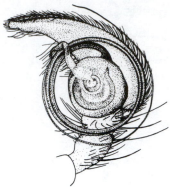

Figure 14. Palp of male *Gnaphosa sericata* from below showing complicated copulatory organ.

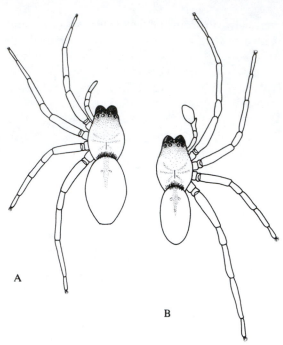

A

B

Figure 17. Comparison of a female spider, A, with a male, B. Note in the male the smaller abdomen, longer legs, and enlarged tarsus of the pedipalp.

Figure 15. Palp of *Gnaphosa sericata* from the side showing an apophysis on the tibia.

paracymbium

patella

Figure 16. Palp of *Pityohyphantes costatus* showing paracymbium, and also an apophysis, on the patella.

segments. That nearest the body is the *coxa*, followed by the *trochanter, femur, patella, tibia, metatarsus* and *tarsus*. Some workers, considering that what is present is a 2-segmented tarsus are using the terms "basitarsus" and "telotarsus" respectively, for the more proximal and more distal segments. The surface of the leg nearest the anterior end of the spider's body is the *prolateral surface,* and that nearest the posterior is the *retrolateral*. In the crab spiders and their relatives the anterior legs may be turned so that the morphologically prolateral surface becomes dorsal; such a leg is called *laterigrade,* a term also used for the sideways locomotion of which these spiders are capable.

The legs are usually covered with hairs and often have spines, or what many now refer to as macrosetae, and bristles, as well as, sometimes, scales. These may all be used in classification. One type of very fine hair, set vertically in conspicuous small sockets, and called *trichobothria* (fig. 18), is of especial use in taxonomy. In some groups the ventral surface of one or more pairs of tarsi (and also in some groups the metatarsi) is provided with a dense brush of short stiff hairs, the *scopula* (fig. 19).

All spiders have at least two claws at the end of each tarsus. In many families an unpaired, median ventral, third claw is present. It is quite small and often difficult to see because of the bristles usually located in this region (fig. 20). In some spiders the bristles are serrated and are considered spurious or accessory claws. In the theridiids a row of such serrated bristles forms a comb along the ventral surface of tarsus IV (fig. 21). Many two-clawed spiders have a dense brush of hairs forming the *claw-tufts* (fig. 23b), but some do not (fig. 22).

Spiders having a *cribellum* (on the abdomen) also have on the dorsal surface, or the dorso-retrolateral edge, of metatarsus IV a series of curved bristles comprising the *calamistrum* (fig. 24).

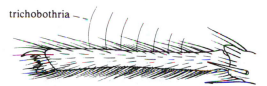

Figure 18. Tarsus of *Coras* showing row of trichobothria.

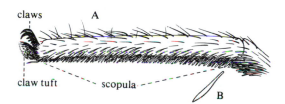

Figure 19. A, Tarsus of *Tibellus* showing scopula. Note also the absence of a lower claw and the presence of claw tufts. B. Scopula hair enlarged.

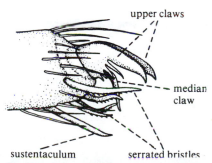

Figure 20. Tip of tarsus IV of *Araneus* showing sustentaculum, bristles and claws.

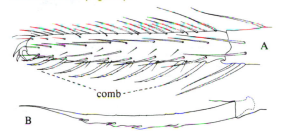

Figure 21. A, *Theridion,* tarsus IV, showing comb; B, Single serrated bristle from comb.

Figure 22. *Agroeca,* tip of tarsus showing two claws without claw tufts.

Figure 23. *Clubiona,* tip of tarsus showing two claws (a) with tufts (b).

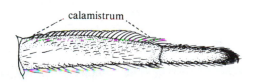

Figure 24. *Dictyna,* showing calamistrum on metatarsus IV.

The pedicel is covered by a sclerotized plate or series of plates called the *lorum*. The shape of these plates is of use in distinguishing certain lycosids and pisaurids. The abdomen is of quite variable size and form. It is most commonly elliptical or oval, but it may be globose, angular, elongate, and either entirely soft, or provided with sclerites, called *scuta* (fig. 25a and b) of variable size.

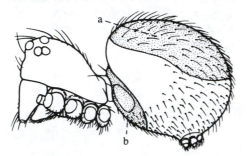

Figure 25. *Pholcomma,* showing abdominal scuta.

In the tarantulas and their kin, as well as in *Hypochilus* there are two pairs of slits on the anterior portion of the venter, each leading into a *lung chamber* (fig. 26). Between the first pair of lung slits in the midline is the slit-like opening of the reproductive system. In some families two pairs of respiratory slits are present, but the second pair leads into tracheal tubes (fig. 27). In other families the second pair of slits is fused to form a single one some distance from the spinnerets (fig. 28), while in most families the single slit leading into the tracheae is situated farther back, immediately in front of the spinnerets, is quite short, and often difficult to see especially if the region is hairy. Usually the anterior lung slits are joined across the venter by an *epigastric furrow*. The region in front of this furrow is commonly more heavily sclerotized than the rest of the venter and may be covered by a distinct plate or scutum in some species. On the dorsum extending back from the anterior end is the *cardiac area* (over the heart) in the mid-line. It is often marked with some special pattern, and in many cases the dorsum as a whole may be marked with a pattern, or *folium* (fig. 29).

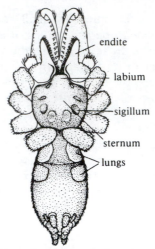

Figure 26. *Atypus* from the ventral aspect, showing two pairs of lungs.

Figure 27. *Ariadna,* abdomen from below, showing pair of tracheal spiracles behind lung slits.

Figure 28. *Neoantistea,* ventral aspect showing spiracle near the middle of the venter.

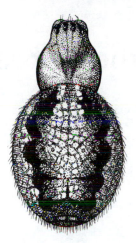

Figure 29. *Zygiella* showing folium on abdomen.

At the posterior end is the anal opening situated on a more or less developed *anal tubercle*. Below this are the *spinnerets,* through which the silk is emitted. A few of our species have only two pairs but the majority by far have three pairs. The thickness, number of segments, as well as the relative lengths of the segments and of the entire spinnerets are all of use in classification. Commonly the medians are the smallest and usually hidden by the others. In a number of families there is present in front of the anterior (or ventral) spinnerets a sieve-like plate, the *cribellum* (fig. 30). The special type of silk emitted from this organ is combed by the calamistrum borne on metatarsus IV. In many of the spiders lacking the cribellum a conical appendage, the *colulus,* lies between the bases of the anterior spinnerets (fig. 31, 32).

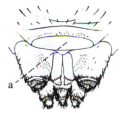

Figure 30. *Hyptiotes,* spinnerets, and showing cribellum at (a).

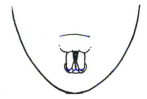

Figure 31. *Loxosceles,* ventral aspect. Stippled structure between the anterior spinnerets is the colulus.

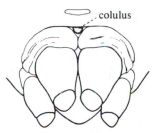

Figure 32. *Xysticus,* spinnerets and colulus.

How to
Recognize the Sexes

Since the keys are based for the most part on mature specimens only, it is necessary to be able to recognize adults and distinguish them from the *spiderlings*. When young both sexes have the palpi simple and leg-like. In most species a male becomes recognizable as such one molt before maturity (in some spiders even two molts before), by the fact that the palpal tarsus is swollen with the developing copulatory organ within. At the final molt these parts become exposed. At the same time may appear other secondary sex characters such as: clasping spines or spurs on the legs or chelicerae, colored scales or hairs on the legs or on the head region, and scuta on the abdomen. The palpi of the female remain simple, but in the majority of spiders at maturity there appears on the ventral side of the abdomen in front of the epigastric furrow a sclerotized plate called the *epigynum* (fig. 33, 34). In the case of some spiders even those with one molt still to go show a very small sclerotized area close to the epigastric furrow. Only experience can enable one to recognize these as being still immature. Likewise there may be difficulty in recognizing mature females in those few groups in which an epigynum is not formed at all.

Although in some families the sexes are alike in size and general appearance, in others there exists considerable sexual dimorphism.

Figure 33. *Leucauge,* venter of female showing position of epigynum between lungs.

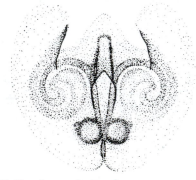

Figure 34. *Gnaphosa sericata,* epigynum.

The males of *Argiope, Nephila, Latrodectus,* and *Misumena* are much smaller than the females and in *Tidarren* very much smaller. Among those spiders which have sclerotized plates, or *scuta,* on the abdomen, the female often has them smaller, or even in many cases lacking entirely. In the Lycosidae and Salticidae the males of many species have special structures which may be displayed to the female during courtship. These generally consist of brushes of hairs on legs and palpi, or colored scales or hairs on the body. Finally, the males of many species are provided with holdfast structures by means of which the females are held during mating. These may be spurs and spines on the legs as in trap-door spiders and orb weavers or the chelicerae may be modified as in *Dictyna* and *Tetragnatha*.

A Word About Spider Venom

So far as known, with the exception of the members of only two small families, all spiders have *poison glands*. These glands open by a pore near the tip of each cheliceral fang. The gland itself is more or less cylindrical, and covered by a layer of spirally arranged muscles (fig. 35) which contract to expel the *venom*. In the majority of our spiders the glands are relatively large, extending back beyond the middle of the cephalothorax as illustrated for *Dolomedes tenebrosus*. In a few groups the glands are relatively larger, but in some they extend barely, or not at all, beyond the bases of the chelicerae. The glands and venom duct as they appear in the black widow are shown in figures 36 and 37.

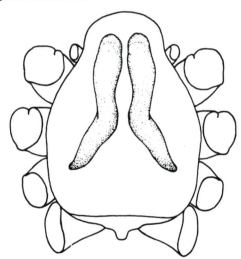

Figure 36. *Latrodectus mactans,* showing location of poison glands in cephalothorax, from above.

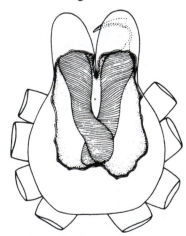

Figure 35. *Dolomedes tenebrosus,* dissection of cephalothorax to show poison glands in position.

The venom is apparently used by spiders to kill their prey, and as a means of defense. The quantity ejected can be controlled by the spider, and may vary with the latter's age and physiological condition, as well as with the degree of irritation to which the spider is subjected. Only a few species, like certain tarantulas and other spiders of the tropics, produce a venom virulent enough to be harmful to man.

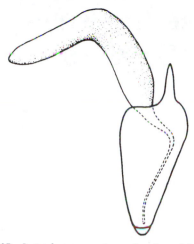

Figure 37. *Latrodectus mactans,* showing right poison gland, its duct and chelicera from the side.

Moreover, most species are too timid and do not attempt to bite even when handled roughly. The bite of our larger wolf spiders is apparently no worse than the sting of a bee or wasp, and that of our tarantulas is hardly more dangerous. Nevertheless, there are some spiders of our region that have been troublesome to man, and produce symptoms of what in the medical literature has been called *arachnidism.*

To begin with, apparently all species of black widows *(Latrodectus)* are known to cause illness. The venom is *neurotoxic,* and while the symptoms vary, the victim usually suffers from a painful rigidity of the abdominal wall muscles. There is usually contraction of the intercostal muscles with a feeling of "tightness" of the chest, and contractions of muscles in the legs. There may also be an increased blood pressure, a rise in body temperature and white blood cell count, and often profuse perspiration, nausea, localized edema, etc. Although relatively few of those bitten die, death may result in from 14 to 32 hours from asphyxia preceded by convulsions. Treatment consists of the application of tincture of iodine or other antiseptic to the site of the bite to prevent infection and the patient is put to bed. Antisera do not appear to be in general use in the United States. To relieve muscle pain an in-travenous injection of 10% calcium gluconate, or of methocarbimol (Robaxin) may be given.

Since 1957 many cases of envenomation by the brown recluse (or "violin" spider), *Loxosceles reclusa,* have been reported. While there may be some slight neurotoxic effects the most obvious symptoms are of a necrotic nature. The actual bite itself may be painless, though sometimes a slight stinging sensation is perceived. The main symptoms appear approximately six to eight hours later. The local reaction includes erythema, tenderness and bleb formation. Within 12 to 24 hours the patient experiences malaise, chills, fever, and nausea. Beginning about the second day, and becoming more noticeable on the sixth or seventh day, skin loss is apparent, followed by extensive necrosis. Ulceration develops and a well defined eschar appears, which is slow to heal. Sometimes there is liver and/or kidney damage associated with hemolysis, and it is this damage which may result in the death of the victim. Prompt administration of corticosteroids has been urged, and some physicians have indicated that the lesions may heal after injection of hydroxyzine hydrochloride, or of phemtolamine. Others recommend immediate excision of all involved tissue to prevent necrosis.

While most reported cases have been due to envenomation by *L. reclusa,* there have been a few cases of bites by *L. deserta* and *L. arizonica.* The symptoms are not as severe as with *L. reclusa.* On the other hand, the symptoms shown by victims bitten by *L. laeta* indicate that this latter species has the most virulent venom. In recent years *L. laeta* has been found in a few isolated localities in the United States, with a rather large number of specimens in one colony found in the Los Angeles area in 1969.

Aside from latrodectism and loxoscelism there are occasional instances of other, less severe symptoms, caused by the bites of various other spiders. Chief among these may be mentioned members of the genus *Chiracanthium* and *Phidippus,* especially in southern California, *P. johnsoni* (i.e., *formosus*).

How to Use the Keys and Study Spiders

As with other animals, spider names consist of two parts; the first is the name of the *genus* and is always capitalized, the second is that of the *species* and always begins with a small (or lower case) letter. Both names are latinized and usually printed in italics. The name of the naturalist who first described and named the spider is often appended (as a third part). It is never italicized and may in some cases be enclosed in parentheses. If so enclosed, it is an indication that the species was originally described as belonging in a genus different from the one in which it is now placed. A species may have been moved to a different genus either because of an error in its original placement, or, more commonly, because more intensive work on a group has shown the need of splitting a genus into two or more genera, with the species in question being included in one of the latter new genera.

Theoretically, spiders can be identified by comparing specimens with the original (or later) full descriptions. But how many of us have the time for such an intensive search, even if we do have access to a complete library of scientific journals, reprints, monographs, and books? Thus it is often expedient to use "keys" which are composed of pertinent and salient features from descriptions. A good key, properly used, is a short cut to identification and saves much time.

Now for an example: If we have in hand three specimens, say a wolf spider, a jumping spider, and a crab spider we find that they can be readily distinguished on the basis of the number of tarsal claws, the arrangement of the eyes, and the manner in which the legs are turned. In the key, characters such as these are arranged in couplets, each half of the couplet bearing the same number but a different letter, as 1a, 1b; 2a, 2b; and so on. The characters given are contrasting, and the student, while examining the specimen, must decide which alternative fits. In addition, on occasion there will be given a character which is possessed by all members of the "a" part of the couplet, even though *some* members of the "b" part may also show this character. Since this is not contrasting for all of both parts the character is given enclosed in parentheses. At the end of each statement of characters is a numeral indicating which couplet is next to be tried, until eventually a couplet line ends in a name, which should be that of the specimen in hand.

Beginning with couplet 1 we find that our three specimens fit the alternative 1b, since the chelicerae project downward and there is only one pair of lungs. We proceed to couplet 7. Here 7b is the correct alternative since none of our specimens possess a cribellum or calamistrum. Moving to couplet 15 we again select the alternative "b" since the characteristic arrangement of spines mentioned in "a" is lacking. That takes us to couplet 16, where we find it necessary to proceed to 21, then 29, and then 30. Here we note that for two-clawed spiders we next go to 31, and for three-clawed to 42. Here then is the first parting of the ways, for our wolf spider having three claws on each tarsus will eventually lead us to couplets 45, 46, 48, 49, and finally be "keyed out" in 50a as a member of the Family Lycosidae. In like manner we will be able to separate the crab spider and the jumping spider in couplet 33, for the latter having eyes in 3 rows will take us to couplet 34, and then be "keyed out" in 35a as a member of the Salticidae. For the crab spider we will eventually end up at couplet 40a in the Thomisidae.

This dichotomous type of key is analogous to a system of roads in which we must always choose one of two forks, and in which no two forks lead to the same place. If we reach a couplet where neither alternative fits, then we have either made a mistake in choice farther back, or we have a representative of a family or genus not included in this book. In order to make the keys more workable the less common genera, and the members of the difficult family Micryphantidae are omitted. The keys are intended to apply only to those spiders which are included in this volume. The reader must be ready for the eventuality that he may have in hand one of the less commonly encountered spiders, which will not fit the keys because it belongs to a genus, or species not included here. It is not to be assumed that a spider which is keyed out to genus necessarily belongs to the species illustrated or described. Only a sampling is included; the various species in a genus may be different in size, pattern or color so that separation is facilitated. Or, on the other hand, they may be similar in size, color and pattern, in which case separation is difficult. It must be remembered that for the most part separation to species (and sometimes to genera) is based upon the structure of the

genitalia and araneologists must make use of these characters, which for the most part are not included in this book. The keys are intended for use with mature specimens only and may not fit juveniles. For example, juveniles of *Scaphiella hespera* do not show the scuta referred to on page 92. For each genus keyed out some indication is given of the number of species that may be found in our area. Likewise, for each family an indication is given of the number of genera in our region.

In many cases members of a family can be immediately recognized on the basis of an easily seen character, as the eye arrangement in Salticidae, Lycosidae, and Oxyopidae, the arrangement of leg spines in the Mimetidae, or of spinnerets in the Hahniidae. Usually, however, the characters are not so easily observed and a good microscope (with lamp) is essential. Specimens are best studied completely immersed in alcohol. In order to make accurate and careful observations at any angle, as well as to make measurements with an ocular micrometer, some device for holding the specimen must be used. A very efficient method is to place a small amount of vaseline on the bottom of the observation dish *before* alcohol is poured in. (It is important that the vaseline be put into a perfectly dry dish, otherwise it will float and be useless.) The specimen can be pushed into just the right amount of vaseline and will be held securely for hours. It can be removed from the vaseline with usually none of the latter adhering to it. Still another method involves placing a layer of washed fine sand into the dish after which the preserving fluid is added. The specimen may be pushed into the sand at any angle, and sand piled up as needed, to maintain the proper position of the specimen.

In measuring the size and spacing of the eyes, one should note that in most cases they are surrounded by black rings. These black areas should not be included in measuring the diameter of eyes. Body length measurements, unless otherwise stated, usually include the chelicerae in front and the spinnerets behind.

Considerable variation in size may be encountered in a given species. This applies particularly to the abdomen which can enlarge in accordance with the amount of food, and in females especially when the eggs are forming. Proportions and relative lengths are more useful than absolute measurements. Colors usually fade in alcohol, so cannot be relied upon, but the pattern of pigmentation is frequently a good character. One should be aware that some species are extremely variable as to color, so that specimens may be found to which the descriptions given do not wholly apply.

Sometimes it is necessary to ascertain the number of teeth on the margins of the cheliceral fang furrow. To facilitate this it may be necessary to remove debris with a fine camel's hair brush, and to pry open a fang with a fine bent dissecting needle.

It is hoped that reference to the illustrations[1] will simplify the task of tracing a specimen through the keys. In the case of those genera which have a number of common species two or more of these are illustrated and described briefly.

In 1905 Nathan Banks published the first key to families and genera of American spiders.

It was then considered that our spiders belonged to 29 families and 245 genera. Modern workers recognize a larger number of families and I have indicated 55 as occurring in our region. For the nine families containing members uncommonly encountered no keys to genera, and no species descriptions, are provided. Of the approximately 550 genera known from the United States perhaps one-fifth belong in the difficult family Micryphantidae, keys to the genera of which are omitted from this book. Two hundred and twenty-three genera are included and 519 species.

To assist the reader in visualizing the position of any spider in the "system" a list is appended giving the placement of spider families, in accordance with the view of the author. Those families whose names are preceded by an asterisk have representatives in our region, i.e., North America north of Mexico, and are included in the key.

1. Concerning the illustrations it should be noted that although all spiders have eight legs these are usually omitted from the drawings. In a few cases the legs of only one side are shown, in some case only the femora are drawn. For the most part, however, in order to conserve space, drawings of only the bodies, or of even just the cephalothorax or abdomen alone are supplied.

Families of Spiders; Order ARANEAE

Orthognatha
  Mesothelae (atypical tarantulas)
    Family   LIPHISTIIDAE Thorell 1869
      [1]ANTRODIAETIDAE   Gertsch
        1940 (folding-door trap-door
        spiders)
      [1]MECICOBOTHRIIDAE
        Holmberg 1882
      [1]ATYPIDAE Thorell 1870 (purse
        web weavers)

  Opisthothelae (typical tarantulas)
      [1]THERAPHOSIDAE   Thorell
        1870 ("ordinary" tarantulas)
      PARATROPIDIDAE   Simon
        1889
      PYCNOTHELIDAE
        Chamberlin 1917
      BARYCHELIDAE Simon 1889
      MIGIDAE Simon 1892
      [1]DIPLURIDAE   Simon   1889
        (funnel web tarantulas)
      [1]CTENIZIDAE   Thorell   1887
        (trap-door spiders)
      ACTINOPODIDAE Simon 1892

Labidognatha
  Hypochiloidea
    Family   GRADUNGULIDAE   Forster
      1955
      [1]HYPOCHILIDAE Marx 1888

Neocribellatae
      [1]FILISTATIDAE Ausserer 1867
      [1]OECOBIIDAE Blackwall 1862
      ERESIDAE C.L. Koch 1850
      [1]DINOPIDAE   C.L.   Koch   1850
        (ogre spiders)
      [1]ULOBORIDAE O.P.-
        Cambridge 1871
      [1]DICTYNIDAE O.P.-
        Cambridge 1871
      [1]AMAUROBIIDAE Thorell 1870
      AMPHINECTIDAE   Forster
        1973
      NEOLANIDAE Forster 1973
      PSECHRIDAE Simon 1890
      STIPHIIDAE Dalmas 1917
      TENGELLIDAE Dahl 1908
      [1]ZOROPSIDAE Bertkau 1882
      ACANTHOCTENIDAE   Simon
        1892

Ecribellatae
  [2]Haplogynae (Primitive hunters and weavers)
    Family   SICARIIDAE Keyserling 1880
      [1]SCYTODIDAE   Blackwall   1852
        (spitting spiders)

1. These families have representatives in North America
north of Mexico, and are included in the key to families.

2. Many modern workers no longer consider these terms
applicable and have discontinued using these names.

[1]LOXOSCELIDAE Simon 1890
[1]DIGUETIDAE Gertsch 1949
[1]PLECTREURIDAE Simon 1893
[1]CAPONIIDAE Simon 1890
[1]OONOPIDAE Simon 1890
TETRABLEMMIDAE O.P.-Cambridge 1873
PACULLIDAE Simon 1894
[1]OCHYROCERATIDAE Fage 1912
[1]LEPTONETIDAE Simon 1890
TELEMIDAE Fage 1913
TEXTRICELLIDAE Hickman 1945
[1]DYSDERIDAE C.L. Koch 1837
[1]SEGESTRIIDAE Simon 1893

[2]Entelogynae
Trionycha
  Higher web weavers
  Family   [1]PHOLCIDAE C.L. Koch 1850 (cellar spiders)
           [1]SYMPHYTOGNATHIDAE Hickman 1931
           [1]THERIDIIDAE Sundevall 1833 (comb-footed spiders)
           NICODAMIDAE Simon 1898
           [1]NESTICIDAE Simon 1894
           HADROTARSIDAE Thorell 1881
           [1]LINYPHIIDAE Blackwall 1859 (line weavers)
           [1]MICRYPHANTIDAE Bertkau 1872 (dwarf spiders)
           [1]THERIDIOSOMATIDAE Simon 1881 (ray spiders)
           [1]ARANEIDAE Simon 1895 ("ordinary" orbweavers)
           [1]TETRAGNATHIDAE Menge 1866 (long-jawed orbweavers)
           [1]AGELENIDAE C.L. Koch 1837 (funnel web weavers)
           ARGYRONETIDAE Thorell 1870 (pond water spiders)

DESIDAE Pocock 1895 (marine spiders)
[1]HAHNIIDAE Bertkau 1878

Three clawed hunters
Family   [1]HERSILIIDAE Thorell 1870
         UROCTEIDAE Thorell 1869
         [1]MIMETIDAE Simon 1890 (pirate spiders)
         ARCHAEIDAE C.L. Koch 1854
         MECYSMAUCHENIIDAE Simon 1895
         [1]ZODARIIDAE Thorell 1881
         PALPIMANIDAE Thorell 1870
         STENOCHILIDAE Thorell 1873
         [1]PISAURIDAE Simon 1890 (nursery-web spiders)
         [1]LYCOSIDAE Sundevall 1833 (wolf spiders)
         [1]OXYOPIDAE Thorell 1870 (lynx spiders)
         SENOCULIDAE Simon 1890
         TOXOPIDAE Hickman 1940

Dionycha (two clawed hunting spiders)
Family   AMMOXENIDAE Simon 1893
         PLATORIDAE Simon 1890
         [1]GNAPHOSIDAE Pocock 1884 (ground spiders)
         [1]PRODIDOMIDAE Simon 1884
         [1]HOMALONYCHIDAE Simon 1893
         CITHAERONIDAE Simon 1893
         [1]CLUBIONIDAE Wagner 1888 (foliage spiders)
         [1]ANYPHAENIDAE Bertkau 1878
         AMAUROBIOIDIDAE Hickman 1949
         [1]ZORIDAE F.O.P.-Cambridge 1893

1. These families have representatives in North America north of Mexico, and are included in the key to families.

2. Many modern workers no longer consider these terms applicable and have discontinued using these names.

'CTENIDAE Keyserling 1876
(running spiders)

'SPARASSIDAE Bertkau 1872
(giant crab spiders)

'SELENOPIDAE Simon 1897

'THOMISIDAE Sundevall 1833
(typical crab spiders)

PHILODROMIDAE Thorell
1870 (running crab spiders)

APHANTOCHILIDAE Thorell
1873

'SALTICIDAE Blackwall 1841
(jumping spiders)

'LYSSOMANIDAE Peckham &
Wheeler 1888

1. These families have representatives in North America
north of Mexico, and are included in the key to families.

2. Many modern workers no longer consider these terms
applicable and have discontinued using these names.

General References

The following represent the few works in English, that give general information about spiders, or assist in further identification of at least some groups. The serious student of spiders will find it necessary to use the many publications scattered for the most part throughout the technical periodicals. Most important here will be the Bulletin of the Museum of Comparative Zoology (at Harvard University) and Psyche, in which most of the contributions by Levi and his students have appeared. Also, the Bulletin of the American Museum of Natural History, and the American Museum Novitates, in which have appeared the contributions of Gertsch, and more recently of Platnick. Besides being included in the Zoological Record the titles of new publications appear in the annual lists put out by the Centre Internationale de Documentation Arachnologique (in Paris). The new worker should consider availing himself of these lists and receiving the Journal of Arachnology by joining the American Arachnological Society. For information write the Membership Secretary, Dr. Norman Platnick at the American Museum of Natural History, New York, N.Y. 10024.

Bristowe, W.S.—1939-1941 The Comity of Spiders. 2 vols. London, 560 p.

Bristowe, W.S.—1958 The World of Spiders. London, 304 p.

Comstock, J.H.—1912 The Spider Book. (Rev. ed. 1940) Garden City, 725 p.

Emerton, J.H.—1902 Common Spiders of the United States. Boston, 235 p.

Fabre, J.H.—1912 The Life of the Spider. New York, 403 p.

Gertsch, W.J.—1949 American Spiders. New York, 285 p.

Kaston, B.J.—1948 Spiders of Connecticut. Hartford, 874 p.

Levi, H.W. & L.R.—1968 Spiders and their Kin. New York, 160 p.

McCook, H.C.—1889-1894 American Spiders and their Spinning Work. 3 vols. Philadelphia, 1254 p.

Savory, T.H.—1928 The Biology of Spiders. London, 376 p.

Warburton, C.—1909 Araneae, in Cambridge Natural History, vol. IV, p. 314-421.

Key to the Orders
of ARACHNIDA

1a Abdomen distinctly segmented externally 3

1b Abdomen not distinctly segmented externally 2

2a Abdomen narrowly joined to the cephalothorax by a petiole. Posterior end of the abdomen provided with small appendages, the spinnerets . . . .
. . . . .page 37 ARANEAE

2b Abdomen fused to the cephalothorax so that the body is of one piece. No appendages at the posterior end. (Throughout the entire region.) (fig. 38). . . . .
. . . . ACARINA

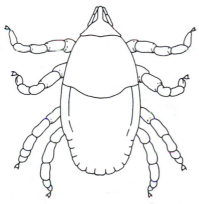

Figure 38. A representative acarine.

3a Abdomen with a wide basal portion (the mesosoma) followed by a narrow distal portion (the metasoma or cauda) terminating in a sting. Pedipalps stout and chelate. (Southeastern States and the Far West.) (fig. 39) . . . .
. . . . SCORPIONIDA

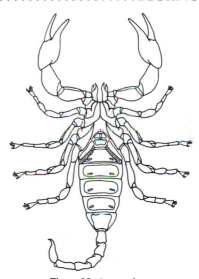

Figure 39. A scorpion.

3b Abdomen not so structured, and without a sting 4

4a Cephalothorax with a movable hood, the cucullus, hiding the mouthparts and front of the head region. (Texas.) (fig. 40) . . . . **RICINULIDA**

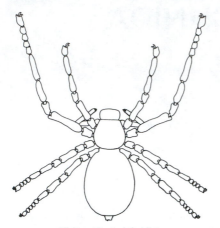

Figure 40. A ricinulid.

4b **Cephalothorax without a cucullus** **5**

5a **Abdomen with a caudal appendage of three or more segments** **6**

5b **Abdomen without a caudal appendage** **8**

6a **Caudal appendage short, at most of three segments. (Texas to California.) (fig. 41).** . . . . **SCHIZOMIDA**

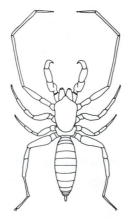

Figure 41. A schizomid.

6b **Caudal appendage long and flagelli-form** **7**

7a **Body length at most 2 mm. Carapace with two or three segments not fused to the anterior ones. Pedipalp pediform, not raptorial. (Texas to California.) (fig. 42)** . . . . **PALPIGRADI**

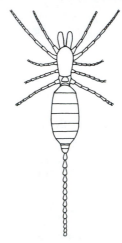

Figure 42. A palpigrade.

7b **Body much larger. Carapace of a single piece covering the entire cephalothorax. Pedipalps heavy and raptorial. (Florida west along the Gulf States to Arizona.) (fig. 43)** . . . . **UROPYGIDA**

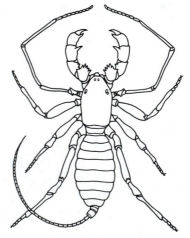

Figure 43. A uropygid.

8a Two or three segments of the carapace not fused to the anterior ones. Chelicerae exceedingly massive. With five T-shaped so-called racquet organs borne under the proximal portion of hind legs. (Florida west along the Gulf States to the Rockies, the Pacific Coast States and adjacent Canada. (fig. 44). . . . . **SOLPUGIDA**

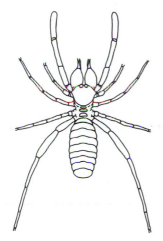

Figure 44. A solpugid.

8b Carapace of one piece covering the entire cephalothorax. Racquet organs absent 9

9a Abdomen narrowly joined to the cephalothorax. Leg I much thinner and very much longer than the others; in life waved about antenna fashion. (Florida to Texas.) (fig. 45). . . . . **AMBLYPYGIDA**

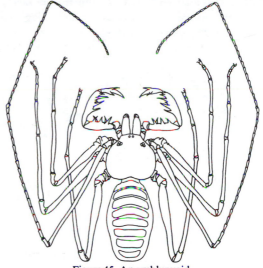

Figure 45. An amblypygid.

9b Abdomen broadly joined to the cephalothorax. Leg I not much, if any, longer or thinner than the others 10

10a Pedipalps long, heavy, and chelate. (Throughout the entire region.) (fig. 46). . . . . **PSEUDOSCORPIONIDA**

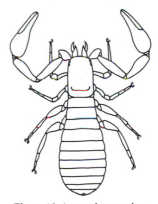

Figure 46. A pseudo scorpion.

10b **Pedipalps not chelate, and not as heavy, in most specimens more leg-like. (Throughout the entire region.) (fig. 47)** . . . . **PHALANGIDA**

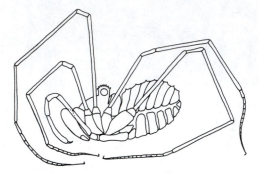

Figure 47. A phalangid.

Pictured Key to the Families of Spiders[1]

1a Chelicerae paraxial, i.e., projecting forward horizontally, and with the fang so articulated as to be movable in a plane more or less parallel to the median plane of the body (fig. 48). With two pairs of lungs. (figs. 49, 50, 65, and 70) . . . . Suborder **Orthognatha** 2

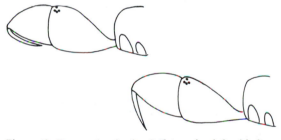

Figure 48. Fang action in the *Orthognatha:* left with fang flexed ("closed"); right with fang extended ("opened").

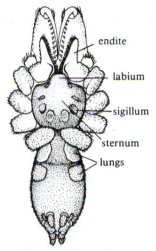

Figure 49. *Atypus niger.*

1. I have not been able to include in this key the members of the Family Symphytognathidae. "This group lies in a twilight zone between families and presents such diluted morphological characters that placement and relationship become uncertain. There are few features that discretely separate them from the Argiopidae [i.e., the Araneidae] on the one hand, or the Theridiidae on the other." (Gertsch 1960). They are all tiny and include the smallest spider ever described, of which a mature male is only 0.5 mm in length. There are known 7 species from 5 genera. Two of the species are from California and Oregon; one from Texas; one from Alabama; two from Florida; and one from the Southeastern States. They are found in forest floor litter.

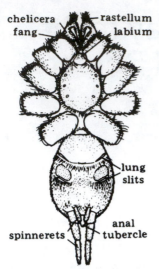

Figure 50. *Antrodiaetus pacificus.*

1b **Chelicerae diaxial, i.e., projecting downward (fig. 51) or in some cases obliquely downward and forward (fig. 52) and with the fangs so articulated as to be movable in a more or less transverse plane (fig. 53). With either two pairs of lungs[2] (fig. 54) or more commonly with only one pair of lungs,[3] and either a single median tracheal spiracle (figs. 55, 56, and 57), or with a pair of spiracles somewhere between the lung slits and the base of the anterior spinnerets (fig. 52). (In Figs. 52, 55 to 57 lung slits are marked ''a'' and tracheal spiracles ''b.'')**. . . . .
. . . . **Suborder Labidognatha 7**

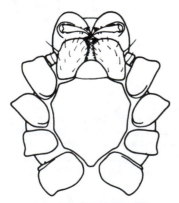

Figure 51. *Ctenium banksi.*

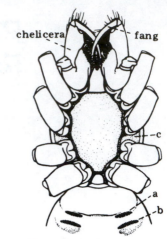

Figure 52. *Dysdera crocata.*

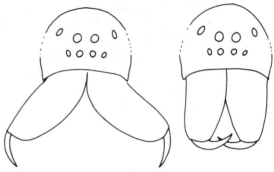

Figure 53. Fang action in the *Labidognatha:* left with fang extended, jaws open; right with fang flexed and jaws closed.

Figure 54. *Hypochilus thorelli; b,* cribellum.

2. If two pairs of lungs are present there will also be a cribellum (fig. 54b), and a calamistrum on metatarsus IV.

3. In the case of a few uncommon spiders, mostly very tiny, no lungs at all, only tracheae.

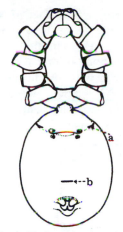

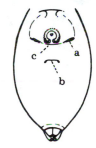

Figure 55. *Scytodes thoracica.*

Figure 56. *Aysha gracilis.*

Figure 57. *Pachygnatha tristriata.*

60). (Four or six spinnerets) . . . .
. . . . The atypical tarantulas 3

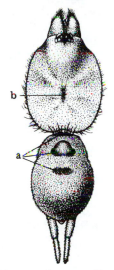

Figure 58. *Antrodiaetus pacificus, male.*

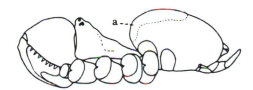

Figure 59. *Atypus niger,* male.

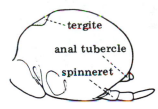

tergite
anal tubercle
spinneret

Figure 60. *Antrodiaetus pacificus,* female, side of abdomen.

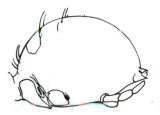

Figure 61. *Antrodiaetus unicolor,* female, side of abdomen.

2a Abdomen with one to three sclerotized tergites (figs. 58, 59, and 60); in some cases with the tergites replaced by transverse rows of bristles (fig. 61). Furrow of cheliceral fang indistinct. Anal tubercle not immediately behind spinnerets, but separated from the spinnerets by a considerable distance (figs. 59, and

2b Abdomen without tergites. Anal tubercle immediately behind the four spinnerets (fig. 62). Furrow of cheliceral fang distinct. (Endites absent or very weakly developed)........................
.... Typical tarantulas 5

Figure 62. *Actinoxia versicolor*, spinnerets and anal tubercle(a).

3a Endites strongly developed and labium fused to sternum (fig. 63). (Six spinnerets. Eight sternal sigilla plainly visible, as in Fig. 63)...........................
.... (p. 65) Family ATYPIDAE

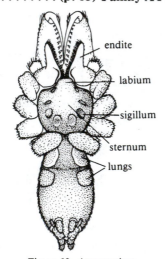

Figure 63. *Atypus niger*.

3b Endites only weakly developed, and labium free. (figs. 64 and 65)........ 4

4a Chelicerae with a row of strong spines, the rastellum, up front (fig. 64). Head region higher than the thoracic region. Labium about as long as wide. Four or six spinnerets, with the last segment of the posterior spinnerets slightly shorter than the penultimate segment, to barely longer than this segment..................
.... (p. 61) Family
ANTRODIAETIDAE

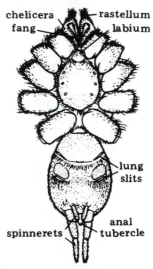

Figure 64. *Antrodiaetus pacificus*.

4b Chelicerae without the rastellum. Carapace fairly flat above, with the head region not higher than the thoracic. Labium much wider than long (fig. 65). Six spinnerets, with the last segment of the hind pair as long as the basal and middle segments together ....
.... (p. 64) Family
MECICOBOTHRIIDAE

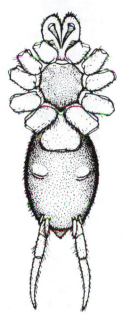

Figure 65. *Hexura picea*, ventral view.

5a Tarsi with a small median (a) as well as two large lateral claws, and without claw tufts. (fig. 66) 6

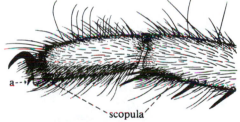

Figure 66. *Myrmeciophila*, foot.

5b Tarsi with only two claws and with claw tufts. (fig. 67) . . . . (p. 66) Family **THERAPHOSIDAE**

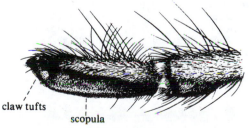

Figure 67. *Dugesiella*, foot.

6a Chelicerae with a rastellum (figs. 68 and 69). Anterior spinnerets not separated by their length, and posteriors with the basal segment as long as, or longer than, the distal and middle together. (Head region much higher than the thoracic region.) . . . . (p. 68) Family **CTENIZIDAE**

Figure 68. *Myrmeciophila*, right chelicera from in front showing rastellum on lobe.

Figure 69. *Actinoxia*, left chelicera from in front, showing rastellum not on lobe.

6b Chelicerae without a rastellum. Anterior spinnerets (a) separated by at least their length, and posteriors (b) very long, with the three segments of about equal length. (fig. 70) . . . . (p. 67) Family **DIPLURIDAE**

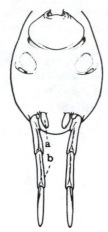

Figure 70. *Euagrus comstocki,* abdomen from below.

7a **With a cribellum in front of the spinnerets[4] (figs. 71, 72, and 73), and a calamistrum on metatarsus IV, varying from just a few bristles to a row the entire length of the metatarsus. (fig. 74)** . . . . . .
. . . . **Section CRIBELLATAE 8**

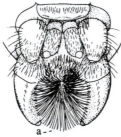

Figure 71. *Oecobius* spinnerets and cribellum. The fringe of hairs on anal tubercle is indicated at a.

Figure 72. *Hyptiotes* spinnerets and showing cribellum at a.

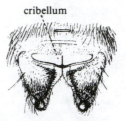

Figure 73. *Amaurobius* spinnerets and showing cribellum.

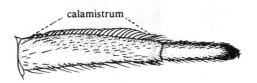

Figure 74. *Dictyna* calamistrum.

7b **Without a cribellum and calamistrum** . . .
. . . . **Section ECRIBELLATAE 15**

8a **With two pairs of lungs. (fig. 75)** . . . .
. . . . **(p. 73) Family HYPOCHILIDAE**

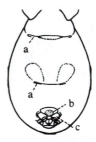

Figure 75. *Hypochilus* thorelli.

4. Note that in males of *Filistata* the cribellum and calamistrum are so rudimentary as to be very difficult to see. The spiders may be recognized and distinguished from ecribellate forms, however, by the fact that the eyes are all on a single raised tubercle, and by the pedipalps being longer than the body.

8b With only one pair of lungs (or none)... 9

9a Anal tubercle large and prominent, two-segmented with a fringe of long hairs (a) (fig. 76). Posterior median eyes triangular or irregular in shape. (Small spiders, 2 to 2.5 mm long with carapace subcircular as in Fig. 77.)............
.... (p. 75) Family **OECOBIIDAE**

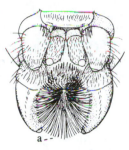

Figure 76. *Oecobius* spinnerets, cribellum and anal tubercle.

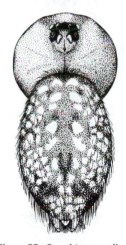

Figure 77. *Oecobius* annulipes.

9b Anal tubercle of the usual type, without a conspicuous fringe of hairs. Posterior median eyes circular.............. 10

10a With the posterior median eyes far forward, and very much the largest (fig. 78). Legs very long, I more than twice, and II almost twice the length of the body. Family **DINOPIDAE**. (With one rare species from Florida and Alabama.)

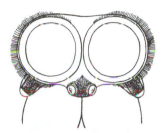

Figure 78. *Dinopis*, eyes from in front.

10b With the posterior median eyes farther back and much smaller. Legs shorter than above......................... 11

11a Tarsi with two claws and with scopulae. Family **ZOROPSIDAE**. (With three uncommon species from Texas and Arizona.)

11b With three claws on tarsi, and without scopulae 12

12a Chelicerae fused together at the base, and each provided distally with a lamella drawn out to a tooth with which the fang forms a kind of chela (fig. 79). Labium fused to the sternum (a). Tracheal spiracle considerably in advance of the spinnerets (fig. 80). Calamistrum short, of only a few setae. (fig. 81)..................
.... (p. 73) Family **FILISTATIDAE**

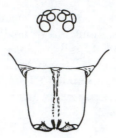

Figure 79. *Filistata*, eyes and chelicerae from in front.

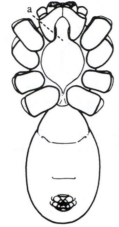

Figure 80. *Filistata*, ventral aspect.

12b Chelicerae not fused at base, and lamella absent. Labium free. Tracheal spiracle in the usual position close to the spinnerets. Calamistrum much longer. (figs. 74, 82, and 83) **13**

13a Tarsi with a dorsal row of trichobothria. Eight eyes all light in color. ....
.... (p. 83) Family
AMAUROBIIDAE

13b Tarsi either without trichobothria, or at most with one. Eight eyes either all dark, or eyes heterogeneous, the anterior medians alone dark; or with only six eyes, the anterior medians lacking **14**

14a Eyes homogeneous, dark, both rows recurved, the posterior row more strongly so, with the laterals of each side farther apart than the two pairs of medians. Metatarsus IV compressed and concave above. (fig. 83) ....
.... (p. 76) Family
ULOBORIDAE

Figure 81. *Filistata*, leg IV showing calamistrum.

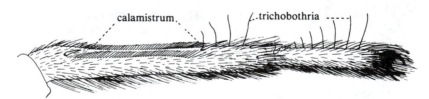

calamistrum   trichobothria

Figure 82. *Callobius bennetti*, leg IV showing calamistrum and trichobothria.

Figure 83. *Hyptiotes*, leg IV showing calamistrum.

14b Eight eyes heterogeneous, the anterior medians alone dark, and at least the anterior row practically straight; or with six eyes pearly white, the anterior medians lacking. Metatarsus IV of the usual shape (not compressed and not concave above).
. . . . (p. 78) Family DICTYNIDAE

15a Tibia and metatarsus I and II with a prolateral row of long spines, in the intervals between which is a row of much shorter spines, curved near their ends, and increasing in length distally. (fig. 84) . . . .
. . . . (p. 175) Family MIMETIDAE

Figure 84. *Mimetus*, metatarsus I showing spination.

15b Tibia and metatarsus I and II either without spines, or spine arrangement not as illustrated in Fig. 84. 16

16a With the chelicerae fused at the base, and each provided distally with a lamella drawn out to a tooth, which with the fang forms a kind of chela. (figs. 85, 86, and 87) 17

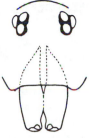

Figure 85. *Spermophora*, eyes and chelicerae.

Figure 86. *Scytodes*, eyes and chelicerae.

Figure 87. *Plectreurys*, face and chelicerae.

16b Chelicerae of the usual type, not fused at the base, but with or without a lamella distally. 21

17a Tarsi long and flexible, with many pseudosegments (fig. 88). Spiracle lacking. Labium broader than long . . . .
. . . . (p. 94) Family PHOLCIDAE

Figure 88. *Pholcus,* flexible tarsus.

17b Tarsi of the usual type, without pseudosegments. Spiracle present and removed from the spinnerets at least one-sixth the distance to epigastric furrow. Labium longer than broad 18

18a Eight eyes in two rows (fig. 87). Labium free. (Three tarsal claws. Spiracular furrow one-third to one-fifth the distance from spinnerets to epigastric furrow.) . . . . (p. 90) Family **PLECTREURIDAE**

18b Six eyes in three diads (as in figs. 86, 89, and 98). Labium fused to sternum 19

19a Anterior row of eyes in a nearly straight line. Carapace only two-thirds as wide as long. Sternum only three-fifths as wide as long. Spiracular furrow conspicuous and one-third the distance from spinnerets to epigastric furrow. (Three tarsal claws. Coxae IV close together.) . . . . (p. 90) Family **DIGUETIDAE**

19b Median diad of eyes far in advance of lateral diads. Carapace much wider and sternum much wider than is the case above. Spiracular furrow not very conspicuous and less than one-third the distance from spinnerets to epigastric furrow. (Colulus large and conspicuous in many) 20

20a Thoracic furrow conspicuous and

longitudinal (fig. 89). Carapace flat above. Tarsi with two claws. Sternum pointed behind. Spiracular furrow one-sixth the distance from spinnerets to epigastric furrow (fig. 90). . . . . (p. 88) Family **LOXOSCELIDAE**

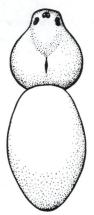

Figure 89. *Loxosceles.*

Figure 90. *Loxosceles,* ventral aspect.

20b Thoracic furrow inconspicuous. Carapace (a) much arched behind (fig. 91). Sternum truncated behind and coxae IV widely separated (fig. 92). Tarsi with three claws. Spiracular furrow (b) one-fourth the distance from spinnerets to epigastric furrow (fig. 92) . . . . (p. 88) Family **SCYTODIDAE**

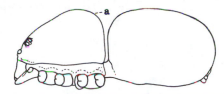

Figure 91. *Scytodes thoracica*, lateral aspect.

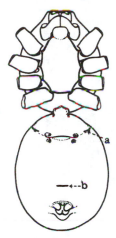

Figure 92. *Scytodes thoracica*, ventral aspect.

21a **With two, four or six eyes**[5] **22**

21b **With eight eyes**[5] **29**

22a **With two eyes (fig. 93) or four. Anterior and median spinnerets in a transverse row** . . . . **Family CAPONIIDAE** **(With three uncommon species in two genera, found under stones in arid and semiarid areas of Texas and the Southwest. These spiders are lungless.)**

Figure 93. Carapace of *Orthonops,* a caponiid.

22b **With six eyes. With the spinnerets in the usual arrangement** **23**

23a **Tracheal spiracle (or spiracles) opening just behind the epigastric furrow** **24**

23b **Tracheal spiracle opening either just in front of spinnerets, or at about the middle of the venter** **26**

24a **Median eyes larger than the laterals (fig. 94A) or else with the anterior laterals contiguous (fig. 94B). Tracheal spiracles inconspicuous and opening in a common transverse furrow. Body length from one to three millimeters. Labium as wide as long. Tarsi with two claws but without tufts** . . . . **(p. 92) Family OONOPIDAE**

Figure 94A. *Orchestina,* eye area.

Figure 94B. *Scaphiella,* eye area from above.

24b **Median eyes not larger than the laterals, and the anterior laterals not contiguous. A pair of conspicuous spiracles (b) opening just behind the lung slits (a), and labium much longer than wide (fig. 95). Body length much more than four milli-**

5. It should be pointed out that occasionally spiders which belong to groups normally possessing eight eyes sometimes show a reduction in number, or of size, of eyes. This is particularly the case with cavernicolous species, and usually the first eyes to be lost are the anterior medians. Obviously the above key cannot take care of these, but only of the usual specimens.

meters. **Tarsi either with three claws or with two claws and tufts. (fig. 96) 25**

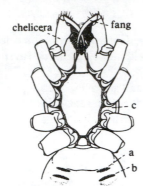

Figure 95. *Dysdera crocata.*

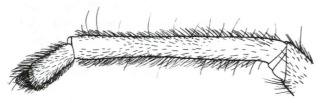

Figure 96. *Dysdera* leg showing claw tufts.

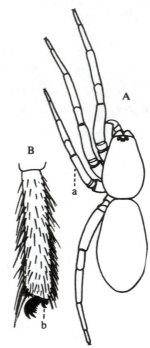

Figure 97. *Ariadna bicolor.* A, dorsal view; B, lateral aspect of tarsus I showing 3 claws.

25a Tarsi with two claws and claw tufts (fig. 96). Leg III directed backward as usual. Sternum with lateral extensions (c) between coxae (fig. 95) . . . .
. . . . (p. 93) Family DYSDERIDAE

25b Tarsi with three claws (fig. 97B). Leg III directed forward together with I and II (fig. 97A). Sternum without the lateral extensions. Face as pictured in Figure 98 . . .
. . . . (p. 93) Family SEGESTRIIDAE

Figure 98. *Ariadna bicolor,* face and chelicerae.

26a Tarsi with three claws borne on a distal extension, the onychium (fig. 99). (Chelicera without boss. Body length usually much less than 2 mm, seldom up to 3 mm.) 28

Figure 99. Onychium of an ochyroceratid.

26b Tarsi with three claws without an onychium 27

27a Eyes in two triads (as in fig. 411). Clypeus much lower than the height of the ocular area. Body length at least 2.5 mm. Female palp with claw. Colulus vestigeal. Anterior spinnerets close together and posterior spinnerets longest .... (p. 164) Family AGELENIDAE (in part)

27b Eyes in three diads, the laterals on each side contiguous. Clypeus much higher than the height of the ocular area. Body length about 1 mm. Female palp without claw. Anterior spinnerets the longest and widely separated. Colulus prominent, as thick as an anterior spinneret......... .... Family TELEMIDAE (With four uncommon species from caves in the far west. These spiders are lungless.)

28a Spiracle just in front of spinnerets. The diad of median eyes much farther back than the other four, which form a recurved row (fig. 100). Pedipalp of female with claw. Chelicera without a lamella. Family LEPTONETIDAE. (This family of small spiders is represented by two genera with 15 species in the Appalachian region, 13 in Texas, and six in California. The spiders are found mostly in caves and under leaf litter, and are not often encountered by the general collector.)

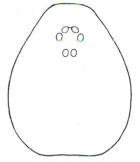

Figure 100. Carapace of *Leptonsta*.

28b Spiracle about midway between epigastric furrow and spinnerets. Eyes in a compact group, the posterior pair right behind the anterior laterals. Pedipalp of female without claw. Chelicera with a lamella (as in figs. 85 and 86). Family OCHYROCERATIDAE. (With a single uncommon species from Florida, and one from the Pacific Northwest.)

29a With six eyes in the front row (fig. 101) .. .. (p. 227) Family SELENOPIDAE

Figure 101. *Selenops*, face.

29b With four or two eyes in the front row 30

30a Tarsi with two claws, with or without claw tufts[6] 31

30b Tarsi with three claws, never with tufts,

6. If tufts are present it can be safely assumed that the tarsus has only two claws.

though spurious claws may be present................................42

31a Tarsal claws without teeth (fig. 102)...32

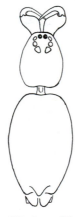

Figure 102. *Homalonychus, tarsus.*

31b Tarsal claws as usual, with teeth......33

32a Chelicerae long and strongly diverging, the margins of the fang furrow toothless, and the fang quite long (fig. 103). Posterior row of eyes strongly procurved, not wider than the anterior, and with the medians triangular.................
..........Family **PRODIDOMIDAE**. (With two rare species; one known from California, and one in another genus from the east coast through the southern States to California.)

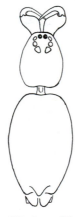

Figure 103. A prodidomid.

32b Chelicerae not so robust, and not extending forward so. Posterior eye row recurved and much wider than the anterior row, the medians circular......

....................(p. 209) Family **HOMALONYCHIDAE**

33a Eyes in three or four rows..........34

33b Eyes in the more common arrangement of two rows......................37

34a Eyes in four rows, the front much the largest (fig. 104)...................
.....(p. 265) Family **LYSSOMANIDAE**

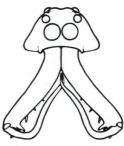

Figure 104. *Lyssomanes* viridis, face and chelicerae of male.

34b Eyes in three rows...............35

35a First row of eyes on a more or less vertical face, the medians much the largest; the second row of two very small, often minute eyes; the third row of two medium-sized eyes (figs. 105, 106, and 107)............................
....(p. 240) Family **SALTICIDAE**

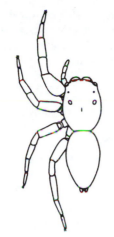

Figure 105. *Metacyrba undata.*

Figure 106. *Phidippus,* carapace from in front.

Figure 107. *Metaphidippus exiguus,* face of male.

35b Front row of eyes not on a vertical face, and the eyes of this row smaller than those of the second **36**

36a First row of two eyes, second with four, and third with two. Anterior laterals much closer to the posterior laterals than to the anterior medians. Retromargin of cheliceral fang furrow with at least three teeth. . . . . **(p. 225) Family CTENIDAE**

36b First row with four eyes; second and third rows each with two (fig. 108). Anterior laterals much closer to anterior medians than to the posterior laterals. Retromargin of cheliceral fang furrow with two teeth . . . . **(p. 224) Family ZORIDAE**

Figure 108. *Zora,* eye area.

37a Tracheal spiracle in advance of the spinnerets at least one-third of the distance between the latter and epigastric furrow (figs. 109 and 110) . . . . **(p. 221) Family ANYPHAENIDAE**

Figure 109. *Aysha,* venter.

Figure 110. *Anyphaena,* venter.

37b Tracheal spiracle in the usual place just in front of the spinnerets **38**

38a At least legs I and II laterigrade, i.e., turned so that the morphologically dorsal surface is posterior and the prolateral surface appears to be the dorsal (figs. 111 and 112). **39**

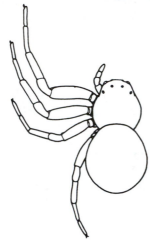

Figure 111. *Synema* bicolor.

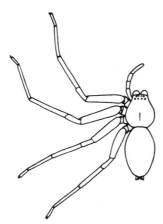

Figure 112. *Heterpoda venatoria*.

38b All legs of the usual prograde type **41**

39a Retromargin of cheliceral fang furrow armed with teeth. Apex of metatarsus with a soft trilobate membrane (a) allowing hyperextension of the tarsus (fig. 113) . . . . (p. 226) Family **SPARASSIDAE**

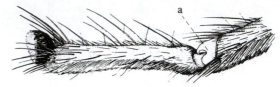

Figure 113. *Heteropoda,* leg I, a. indicating the trilobate membrane between metatarsus and tarsus.

39b Chelicerae with retromargin of fang furrow smooth, the promargin alone with at most one or two teeth. Apex of metatarsus sclerotized (as is usual) so that hyperextension of the tarsus is impossible **40**

40a Colulus present (fig. 114). Hair over body simple and erect. Legs I and II much longer and stouter than III and IV. Claw tufts lacking, or composed of simple hairs, and tarsi I and II not scopulate (fig. 115). Promargin of cheliceral fang furrow unarmed. . . . . (p. 227) Family **THOMISIDAE**

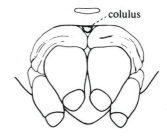

Figure 114. *Xysticus* spinnerets and colulus.

Figure 115. *Misumenoides,* leg showing scopula lacking and claw tufts simple and sparse.

40b Colulus absent. Hair over body feathery or scaly, and prone. All legs about the same length or leg II alone much longer. Claw tufts composed of spatulate hairs and tarsi I and II scopulate (fig. 116). Promargin with one or two teeth . . . .
. . . (p. 235) Family **PHILODROMIDAE**

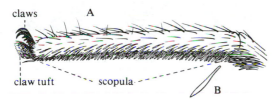

Figure 116. *Tibellus,* tarsus showing scopula and claw tufts, A; and spatulate hair, B.

41a Anterior spinnerets conical, contiguous or almost so, and not more heavily sclerotized than the posterior (fig. 117). Eyes homogeneous or almost so (with few exceptions). Endites without a transverse or oblique depression (except in *Micaria*).
. . . . (p. 210) Family **Clubionidae**

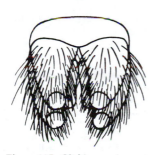

Figure 117. *Clubiona spinnerets.*

41b Anterior spinnerets cylindrical, longer and more heavily sclerotized than the posterior, and separated by a distance about equal to the diameter of one (fig. 118). Eyes distinctly heterogeneous, the anterior medians alone dark; the posterior medians often oblique, oval, or triangular. Endites with an oblique depression (a) on the ventral face (fig. 119) . . . .
. . . . . . (p. 200) Family **GNAPHOSIDAE**

Figure 118. *Gnaphosa* spinnerets.

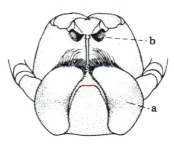

Figure 119. *Gnaphosa,* showing transverse depression on endites (a).

42a The six spinnerets in a more or less transverse row (fig. 120). Tracheal spiracle removed from the spinnerets at least one-third of the distance to epigastric furrow . . . .
. . . . (p. 174) Family **HAHNIIDAE**

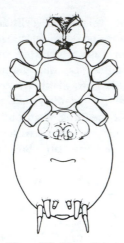

Figure 120. *Neoantistea.*

42b Spinnerets not so placed, but of the usual arrangement. Tracheal spiracle in the usual place in front of the spinnerets (few exceptions) 43

43a The median and posterior spinnerets considerably reduced so that it may appear as though only one pair is present, the anterior (fig. 121). Family ZODARIIDAE. (With four uncommon species in one genus in California, and a very rare one in another genus in Pennsylvania.)

Figure 121. *Lutica,* a zodariid, spinnerets from above.

43b Posterior spinnerets not so reduced; rather, as prominent as the anterior or more so 44

44a Posterior spinnerets exceptionally long, with the apical segment as long as the abdomen, or longer (fig. 122). Colulus prominent. (Legs very long. Body very flat, but with the head region much elevated above the thoracic.) Family HERSILIIDAE (With one uncommon species from south Texas.)

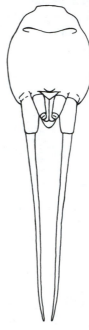

Figure 122. *Tama,* a hersiliid, ventral aspect of abdomen showing long spinnerets and colulus.

44b Posterior spinnerets very much shorter 45

45a Eye group hexagonal, the posterior row procurved, the anterior row recurved, with the clypeus high (fig. 123, 506, or 509). (Abdomen pointed behind and legs with very prominent spines)........... .... (p. 197) Family OXYOPIDAE

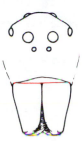

Figure 123. *Oxyopes*, face and chelicera.

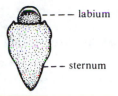

Figure 125. *Tetragnatha*, sternum and labium showing rebordered front edge of latter.

45b Eye group not forming a hexagon, and clypeus much lower 46

46a With tarsus IV in most specimens provided for at least one-sixth its length from the distal end with a ventral row of 6 to 10 serrated bristles, forming a comb (fig. 124), which may be poorly developed in males. Spiders hanging in an inverted position in irregular mesh webs 47

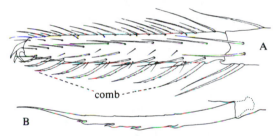

Figure 124. *Theridion*, tarsus IV showing comb of serrated bristles.

46b Tarsus IV without such a comb[7] 48

47a Labium rebordered as in Figure 125. Comb bristles not longer than bristles on the dorsal side of tarsus IV. Margins of cheliceral fang furrow toothed. . . . .
. . . . (p. 115) Family NESTICIDAE

47b Comb bristles longer than the bristles on the dorsal side of tarsus IV (fig. 124). Labium not rebordered (few exceptions). Margins of cheliceral fang furrow without teeth in most species (very few exceptions)
. . . . (p. 97) Family THERIDIIDAE

48a Tarsi with trichobothria (figs. 126 and 127). Lip not rebordered 49

Figure 126. *Lycosa*, tarsus showing trichobothria.

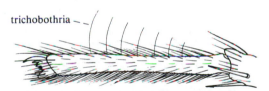

Figure 127. *Coras*, tarsus showing trichobothria.

48b Tarsi without trichobothria. Labium rebordered (fig. 125) 51

49a Tarsi with a single row of trichobothria, which in many species increase in length toward the distal end, as in Figure 127.

7. A few small theridiids would key out here, since they lack the comb characteristic of their family.

Trochanters not notched. For the most part living in sheet webs with a funnel, over which they run rapidly in an upright position . . . .
(p. 164) Family AGELENIDAE (in part)

49b Tarsi with trichobothria numerous, but irregularly distributed (fig. 126). All trochanters with a curved notch (a) along the distal edge of the ventral side (fig. 128) 50

Figure 128. *Lycosa*, ventral aspect of trochanter.

50a Posterior row of eyes so strongly recurved that it may be considered to form two rows (figs. 129, 130, 131, and 132). Median claw smooth or with a single tooth. Anterior piece of lorum (a) rounded behind and fitting into a notch of the posterior piece (b) (fig. 133). Egg sac carried attached to spinnerets (fig. 459) and young carried on mother's back (fig. 460) . . . .
. . . . (p. 181) Family LYCOSIDAE

Figure 129. *Trabea*, eyes from in front.

Figure 130. *Pardosa*, eyes from in front.

Figure 131. *Trochosa*.

Figure 132. *Lycosa*.

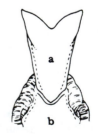

Figure 133. *Lycosa*, lorum of pedicel.

50b Posterior row of eyes not forming two distinct rows, but only slightly recurved (figs. 134 and 135). Median claw with two or three teeth. Anterior piece of lorum (a) with a notch (c) into which the posterior piece (b) fits, or a transverse suture (c) between the two pieces (figs. 136 and 137). Egg sac held under cephalothorax (fig. 4); young not carried about by mother . . . . (p. 177) Family **PISAURIDAE**

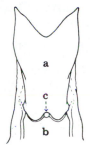

Figure 137. *Dolomedes*, lorum of pedicel.

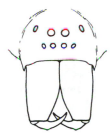

Figure 134. *Pisaurina*, eyes from in front.

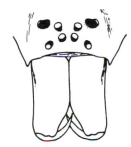

Figure 135. *Pelopatis*, eyes from in front.

Figure 136. *Pisaurina*, lorum of pedicel.

51a Clypeus (a) in most lower than the height of the median ocular area (fig. 138). Eyes homogeneous. Chelicera with a boss, though rudimentary in some. (For the most part weavers of orb webs.) 52

Figure 138. *Araneus*, face and chelicerae.

51b Clypeus usually as high as, or more commonly higher than, height of the median ocular area (fig. 139). Eyes heterogeneous. Chelicera without a boss. The majority do not weave orb webs . . 53

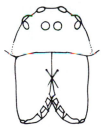

Figure 139. *Pityohyphantes*, face and chelicerae.

52a Epigastric furrow between lung slits procurved (fig. 140). Rudimentary boss on chelicerae. In most specimens the chelicerae are large and powerful (fig. 141). Femora with trichobothria . . . . . (p. 160) Family **TETRAGNATHIDAE**

Figure 140. *Tetragnatha,* venter showing procurved epigastric furrow.

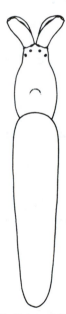

Figure 141. *Tetragnatha* straminea.

52b Epigastric furrow nearly straight. Boss conspicuously present on chelicerae in most (fig. 142), though rudimentary in some. Femora without trichobothria, or if present then femur IV provided with a double fringe of hairs on the prolateral surface (fig. 337A). . . . . (p. 130) Family **ARANEIDAE**

Figure 142. *Eustala,* chelicera from side showing boss.

53a Sternum broadly truncate behind. Femur I (a) about three times as thick as IV (fig. 143). Legs without spines. Weave modified orb webs (fig. 144). (Palp of female without claw, and chelicera without a stridulating area.) . . . . (p. 128) Family **THERIDIOSOMATIDAE**

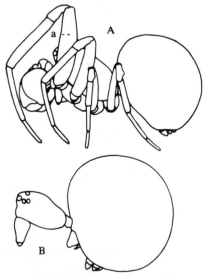

Figure 143. *Theridiosoma.* A, male; B, female.

Figure 144. *Theridiosoma* snare.

53b **Without this combination of characters. Web is not an orb, but an irregular net, or a modified sheet. (Lateral face of chelicera in many species with a stridulating area (a) as in (Fig. 145.) 54**

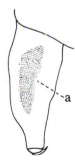

Figure 145. *Meioneta micaria,* chelicera from side showing stridulating area (a).

54a Tibia of male pedipalp without apophyses (though the tibia may be dilated distally). (fig. 146). Palp of female in most species with a claw at the end of the tarsus. Tibia IV in most species with two dorsal spines, or if only one spine is present then there is one short spine on metatarsi I and II (fig. 147) . . . . (p. 115) Family LINYPHIIDAE

54b Tibia of male pedipalp in most species with at least one apophysis (fig. 148). Palp of female without a claw at end of tarsus. Tibia IV with a single dorsal spine or bristle, and with the metatarsi spineless, or all spines lacking altogether . . . (p. 128) Family MICRYPHANTIDAE

Figure 148. *Ceraticelus* palp, with tibial apophysis at a.

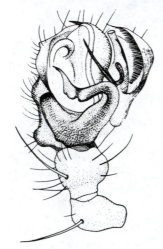

Figure 146. *Lepthyphantes* palp of male.

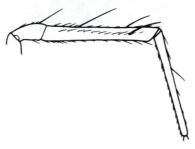

Figure 147. *Lepthyphantes,* patella, tibia and metatarsus of leg IV.

Descriptions of Families and Keys to the More Common Genera

SUBORDER ORTHOGNATHA

This suborder, often referred to as the Mygalomorphae, includes the tarantulas and their allies, most of which are heavily bodied and have stout legs. The poison glands lie entirely within the chelicerae, which are very large and powerful. The majority of species are tropical, or subtropical, but a few are not uncommon in the United States, particularly in the South and West.

FAMILY ANTRODIAETIDAE

Foldingdoor Trap-door Spiders

At least some of the members of this family close their burrows with a door consisting of two semicircular halves meeting on the midline of the opening. This is a small family, containing three genera, with representatives more commonly found in the more southern and western portions of the continent. The males tend to wander at night (at least in California) after the ground has been soaked by heavy rains, especially from November to February. After the rains the females may be found out-

side usually at night repairing the doors to the burrows and laying down fresh silk.

1a With only four spinnerets, the anterior laterals absent (Pedipalpal patella of male shorter than tibia) . . . . *Antrodiaetus* (11 species)

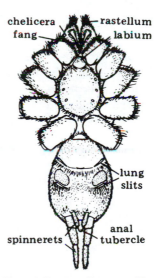

Figure 149. *Antrodiaetus pacificus.*

61

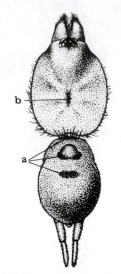

Figure 150. *Antrodiaetus pacificus.* male.

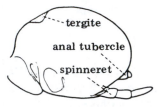

Figure 151. *Antrodiaetus pacificus,* female, side of abdomen.

Figures 149, 150, and 151. *Antrodiaetus pacificus* (Simon)

The cephalothorax and legs are brown, the abdomen brownish gray. The male has three abdominal tergites, and the female one. Length of female 13 mm; of male 11 mm.

    Alaska south to California and east to Idaho.

Figure 152. *Antrodiaetus unicolor,* female, side of abdomen.

Figure 152. *Antrodiaetus unicolor* (Hentz)

The cephalothorax and legs are dark chestnut brown, and the abdomen is light purplish brown, with the venter paler. While the male shows three distinct tergites on the abdomen the female shows only one, which is presumably the homolog of the middle one in the male. The anterior and posterior tergites are represented by a transverse row of bristles. The tergites themselves have bristles arising from them too. Length of female 20 mm; of male 17 mm.

    New York southwest to Arkansas and south to Georgia and Alabama.

1b With six spinnerets, the anterior laterals present although they may be very small 2

2a Thoracic groove longitudinal. Anterior lateral spinnerets more slender than the anterior medians, and of a single segment (difficult to see in some) (fig. 153A). Posterior median and anterior median eyes subequal in size. Chelicera of male with a large horn-like apophysis projecting anterodorsally (fig. 153B). Pedipalpal patella much shorter than tibia . . . . *Atypoides* (3 species)

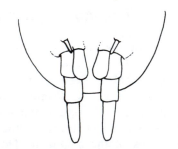

Figure 153A. *Atypoides, A, spinnerets.*

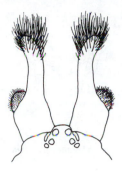

Figure 153B. Head region of male showing cheliceral apophyses.

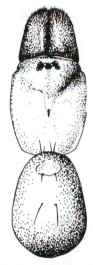

Figure 154. *Atypoides riversi*, female.

Figure 154. *Atypoides riversi* O.P.-Cambridge

The male has three abdominal tergites and the female has one. These are brown in color, but the rest of the abdomen is purplish brown. The cephalothorax and legs are dark chestnut brown. The posterior lateral spinnerets have three segments, the posterior medians one. Anterior lateral spinnerets at least one-third as long as the posterior medians. Length of female 17 to 18 mm; of male 13 to 15 mm.

The spider builds a burrow with the entrance fashioned as a stiffened turret or collapsible collar, but no door.

California.

Atypoides gertschi Coyle

The male has two abdominal tergites, and the female one. The color of the abdomen is grayish yellow or pale gray. The anterior lateral spinnerets are at most one-fifth as long as the posterior medians, or in some are so small as to be difficult to see. Length of female 20 to 28 mm; of male 18 to 21 mm.

Oregon and northern California.

2b Thoracic groove an irregular pit from transverse to irregular, or else absent altogether. Anterior lateral spinnerets almost as thick as the anterior medians, and of two segments (fig. 155). Diameter of posterior median eyes hardly half that of the anterior medians. The patella of the male pedipalp is greatly elongated so that the entire pedipalp is about as long as leg I. Chelicera of male lacks the horn-like apophysis........................ ....*Aliatypus* (11 species)

Figure 155. *Aliatypus, spinnerets,*

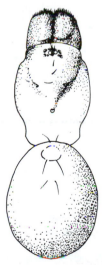

Figure 156. *Aliatypus californicus*, female.

Figure 156. *Aliatypus californicus* (Banks)

The thoracic groove is a deep pit. The cephalothorax and legs are yellow to light brown. The abdomen is brownish gray, lighter on the venter. Length of female up to 23 mm; of male 12 mm.

This species has been reported constructing a burrow with a wafer door.

California.

Aliatypus thompsoni Coyle
Similar to *californicus* but smaller. Also, the thoracic groove is nearly always absent, or a very small depression. Length of female 8.8 to 14.4 mm; of male 6.1 to 8.9 mm.

California.

FAMILY MECICOBOTHRIIDAE

Sheet-web Weaving Atypical "Tarantulas"

There are three genera, one in the Eastern States. These spiders live under leaves, loose bark, pieces of wood and trash on the ground in wooded areas. A sheet-like web is constructed, somewhat like that of the Dipluridae. As a matter of fact the members of these two families show a decided convergence of structure and habit.

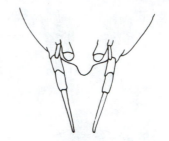

Figure 157. *Megahexura, spinnerets.*

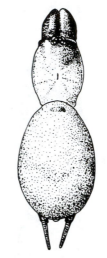

Figure 158. *Megahexura fulva.*

Figure 158. *Megahexura fulva* (Chamberlin)

The body and legs are a light brown color. Length (including spinnerets) of female 15 to 18 mm; of male 10 to 13 mm.

California.

1a Abdomen with two sclerotized tergites. Diameter of anterior lateral eyes hardly three times that of the anterior medians. Metatarsus I with three pairs of ventral spines. Anterior lateral spinnerets two-segmented (fig. 157). Length of hind spinnerets less than half the length of abdomen and (at least in the female) less than the width of the abdomen . . . .
. . . . *Megahexura* (1 species)

1b Abdomen with at most only one tergite. Diameter of anterior lateral eyes more than four times that of anterior medians. Metatarsus I with only two pairs of ventral spines. Anterior lateral spinnerets of a single segment (fig. 159). Length of hind spinnerets about four-fifths that of the

abdomen's length, and greater than its width. . . . .
. . . . *Hexura* (1 species)

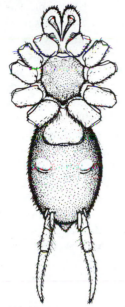

Figure 159. *Hexura picea,* ventral view.

Figure 159. *Hexura picea* Simon

The cephalothorax and legs are brownish, and the abdomen is purplish. Length of female about 12 mm including the long spinnerets; of male about 8.5 mm.

Oregon and Washington.

FAMILY ATYPIDAE

Purse-web Spiders

The webs are silk-lined tubes in the ground, and extend eight to 10 inches up the side of a tree trunk, or along the surface of the ground. When that portion of the tube above ground is disturbed by the passage of an insect over it the spider rushes to the place the insect is and captures it by biting through the webbing. The tube is then slit and the prey dragged in. This is a small family represented in our region by the single genus *Atypus,* with four species.

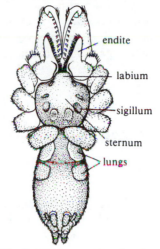

Figure 160. *Atypus niger,* female.

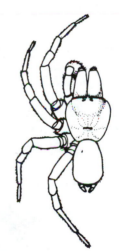

Figure 161. *Atypus niger,* male.

Figures 160 and 161. *Atypus niger* (Hentz)

The body is dark brown to black with the legs quite stout. The male has a single abdominal tergite, the female none. Length of female 17 mm; of male 10 mm.

Ontario and New England south to North Carolina and west to Wisconsin and Kansas.

Atypus bicolor (Lucas)
Similar in form and color to *niger,* but having the legs yellow to reddish. Length of female up to 25 mm; of male 14.5 mm.

New England south to Florida and west to Louisiana. Chiefly in mesophytic woods; less common than *niger* in northern parts of its range.

FAMILY THERAPHOSIDAE

Tarantulas

These are our largest spiders, and because of their great size and their hairiness (fig. 162) often attract attention and are much feared. While it is true that some of the South American tarantulas have a deadly bite, experiments on our species have shown their bites to be relatively no more harmful than a bee sting. The spiders are nocturnal for the most part, hiding during the day inside of natural cavities in the ground, in abandoned rodent tunnels, or similar places, the upper portions of which they line with silk. At times deep burrows are made. Males, at least in southern California, are often found wandering from July to November.

Many tarantulas when disturbed have the habit of using their hind legs to scrape hairs off the posterior part of the abdomen, so that the latter often shows a bald spot. The hairs may be urticarial to a greater or lesser degree, varying with the species, and may serve to ward off vertebrate predators. Females live for many years (up to 35?) and usually molt at approximately yearly intervals. There are five genera with about 30 species in the family, which is also known under the name Aviculariidae, and they occur most commonly in our southwestern States. They resemble each other and are difficult to identify.

1a The prolateral surface of coxa I is provided with setae which are conspicuously thickened basally (fig. 163) . . . .
. . . . *Dugesiella* (8 species)

Figure 163. *Dugesiella,* prolateral surface of coxa I.

Figure 162. A *theraphosid.*

Dugesiella hentzi (Girard)
The cephalothorax and legs are dark brown, with reddish golden hairs on the carapace. The abdomen is brownish black. Length of female 44 to 58 mm; of male 38 to 52 mm.

Southern Missouri and Kansas south through Oklahoma, Arkansas and Louisiana.

1b **The prolateral surface of coxa I is provided with setae that are not thickened much basally (fig. 164)** . . . . *Aphonopelma* **(20 species)**

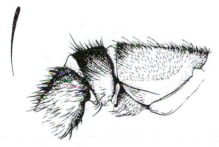

Figure 164. *Aphonopelma,* prolateral surface of coxa I.

Aphonopelma eutylenum (Chamberlin)
Mostly blackish to dark brown. Length of female 43 mm; of male 40 mm.

California.

Three other species also found in the New Mexico, Arizona, and southern California area include: A. *reversum* Chamberlin, of about the same size as *eutylenum;* A. *baileyi* Chamberlin, which is somewhat smaller; and A. *chalcodes* Chamberlin, which is somewhat larger.

FAMILY DIPLURIDAE

Sheet- or Funnel-web Building Tarantulas

At least some of the members of this family, e.g., *Euagrus,* spin a funnel in a crevice, under rocks, or in thick vegetable growth, and continue the silk as a large sheet over the ground. Others, e.g., *Calisoga,* do not. This is a small family of three genera, which, except for one species, are restricted to the West and Southwest.

1a **Thoracic fovea (a) a circular pit (fig. 165). Cephalothorax flat above with head region not higher than the thoracic region. Tarsi without scopulae. Anterior spinnerets separated by three times the diameter of one (fig. 166)** . . . . *Euagrus* **(6 species)**

Figure 165. *Euagrus comstocki,* cephalothorax, from above.

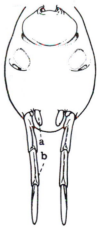

Figure 166. *Euagrus comstocki,* abdomen from below.

Figure 167. *Euagrus comstocki,* female.

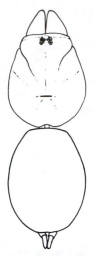

Figure 168. *Calisoga longitarsus,* female.

Figure 167. *Euagrus comstocki* Gertsch

The cephalothorax and legs are brown, the abdomen tan to dirty yellow. Length of female 15 mm; of male 13 mm.

Texas.

Euagrus ritaensis Chamberlin & Ivie

Similar to *comstocki.* Length of female 10 mm; of male 8 mm.

Arizona.

1b **Thoracic groove transverse. Cephalothorax with head region much higher than the thoracic region. Tarsi scopulate. Anterior spinnerets separated by only twice the diameter of one. . . . .**
. . . . *Calisoga* **(2 species)**

Figure 168. *Calisoga longitarsus* (Simon)

This has long been known under the name Brachythele. Besides the scopulae on the tarsi there are scopulae on metatarsi I and II, and near the distal end of metatarsus III. These scopulae are not as thick as in tarantulas, and of course there is a median claw present, but in general this species resembles a tarantula without the long hairs on the abdomen. A burrow is constructed, the upper fourth of which is lined with silk, which spreads out from the burrow's mouth. There is no door. Length of female 25 to 46 mm; of male 15 mm.

California and Nevada.

FAMILY CTENIZIDAE

Trap-door Spiders

These are the spiders, known from our southern and western states which have developed the habit of digging in the ground tubular burrows, the walls of which are lined with silk, and the entrances to which are closed by a single-pieced lid (fig. 169), hinged with silk at one side. The lids are made to fit so snugly and are so often camouflaged with bits of vegetation from the surrounding area, that they commonly require the closest scrutiny on the part of the collector if he desires to locate them. The lids may be thin flaps of silk, of the *wafer* type, which do not fit as snugly, or as in the *cork* type thick with included soil, and beveled to fit the opening. They are never of the folding type characteristic of the Family Antrodiaetidae. In those species with cork doors the spider is capable of holding the door shut with surprising strength. This is the most widespread of the orthognath families, with 10 genera in our region.

Some species build tunnels with a branch, so the entire structure is shaped like a "Y," and others with side chambers, which themselves are closed off from the main tunnel by means of hinged doors (as in *Myrmeciophila*). In southern California the males tend to wander after the ground has been soaked by heavy rains, especially from November to February.

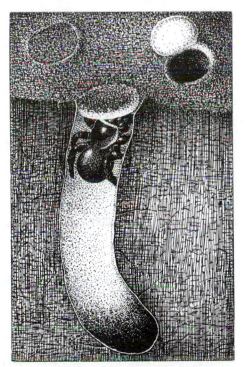

Figure 169. *Bothriocyrtum* in its nest.

1a **Abdomen truncated and discoidal caudally, and with numerous longitudinal grooves (fig. 170).** . . . . *Cyclocosmia* **(2 species)**

Figure 170. *Cyclocosmia*.

GENUS CYCLOCOSMIA

The cephalothorax and legs are brown, as is also the anterior portion of the abdomen, but the truncated part is black. These peculiarly shaped spiders live in ravines in the Southeastern States. Most specimens have been found in deep vertical burrows in river banks.

Cyclocosmia truncata (Hentz)
In this species there are 24 longitudinal ribs on each side of the abdomen. Length of female 30 mm; of male 19 mm.

Tennessee, north Georgia and north Alabama.

Cyclocosmia torreya Gertsch & Platnick
In this species the abdomen has only 22 ribs on each side. Length of female 33 mm; of male 18 mm.

North Florida and south Georgia.

1b **Abdomen of the usual type and without longitudinal grooves** 2

2a **Tarsi I and II with lateral rows of short stout spines. Anterior spinnerets at least three-fifths as long as the basal segment**

of the posterior spinnerets. Carapace about as broad as long 3

2b Tarsi I and II without the lateral rows of short spines. Anterior spinnerets at most only about half as long as the basal segment of the posterior spinnerets. Carapace definitely longer than broad . . 4

3a Tibia III with a deep shiny depression (a) (fig. 171) . . . .
. . . . *Ummidia* (11 species)

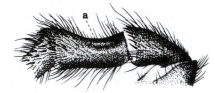

Figure 171. *Ummidia*, showing depression on tibia III.

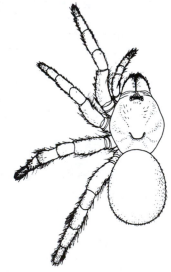

Figure 172. *Ummidia audouini,* dorsal view of female.

Figure 172. *Ummidia audouini* (Lucas)

The cephalothorax and legs are chestnut brown and the abdomen black. Length of female 28 mm; of male about 15 mm.

This is the common trap-door spider of the southeastern states, building its tunnels in the sides of steep banks. The burrows are shallow, generally five to eight inches deep, and often inclined toward the horizontal, rather than being strictly vertical. The door is thick, of the cork type.

Virginia south to Florida and west to the Gulf States, Oklahoma and Illinois.

3b Tibia III of the usual type, without a depression . . . .
. . . . *Bothriocyrtum* (1 species)

Figure 173. *Bothriocyrtum*, cephalothorax from above.

Figure 173. *Bothriocyrtum californicum* (O.P.-Cambridge)

Cephalothorax brown, darker in male. Abdomen brownish to yellow-gray, lighter in male. Length of female about 28 to 33 mm; of male 18 to 24 mm. It builds mostly on sunny hillsides and has a thick cork lid.

This is the common trap-door spider of southern California.

4a With the rastellum borne on a stout lobe or process at inner apex of chelicera (fig. 174). (Posterior sigilla of sternum closer to each other than to the edge of sternum) . . . .
. . . . *Myrmeciophila* (4 species)

Figure 174. *Myrmeciophila*, right chelicera showing rastellum borne on a lobe.

Figure 175. *Myrmeciophila* cephalothorax from above.

Figure 175. *Myrmeciophila fluviatilis* (Hentz)

The cephalothorax and legs are light brown, the abdomen gray. Length 17 to 23 mm. The door is of the wafer type and the tunnel may have branches.

Tennessee and North Carolina, south to Florida.

4b **Rastellum not borne on such a lobe (fig. 176)** **5**

Figure 176. *Actinoxia*, left chelicera from in front showing rastellum not on a lobe.

5a **Coxa of pedipalp provided with spinules which are scattered from base to apex. Posterior sigilla of sternum relatively large and nearer to each other than to the edge of the sternum (fig. 177A)** . . . . *Actinoxia* (3 species)

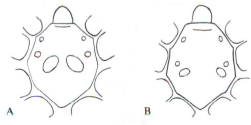

Figure 177. Sternum showing sigilla; A, of *Actinoxia;* B, of *Aptostichus.*

Figure 178. *Actinoxia versicolor.*

Figure 178. *Actinoxia versicolor* (Simon)

The cephalothorax is brown; the abdomen is pale yellow with a pattern of gray chevrons. Length of female about 25 to 26 mm; of male about 13 to 17 mm. The door is of the wafer type.

Central California.

5b **Coxa of pedipalp with spinules which are**

limited to at most the basal half (fig. 179). Posterior sigilla of sternum relatively small and farther from each other than they are from the sternal border (fig. 177B). . . . .
. . . . *Aptostichus* (5 species)

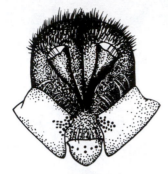

Figure 179. *Aptostichus,* mouthparts.

GENUS APTOSTICHUS

Some species seem to bs restricted to sand dunes and the desert, while others build their burrows in dirt. There are many species (perhaps 20) that have been collected, but not yet described.

Figure 180. *Aptostichus stanfordianus* male.

Figure 180. *Aptostichus stanfordianus* Smith

Carapace light brown, darker along the cephalic groove. The abdomen is yellowish brown, with a series of darker brown blotches above, and paler beneath. Length of female 18 to 23 mm; of male 13 to 15 mm. The door is of the wafer type.

California.

SUBORDER LABIDOGNATHA

This group, often called Araneomorphae, includes what are usually referred to with the unfortunate term "True spiders," comprising by far the majority of those apt to be encountered in the United States and Canada. Their average size is much smaller than that of the Orthognatha, and many species are quite minute, less than 2 mm body length. The endites are always provided with lobes which assist in the chewing of food. Based upon the presence or absence of a cribellum and calamistrum the spiders are placed in the Section Cribellatae, or Ecribellatae, respectively.

Section Cribellatae

These spiders catch their prey by means of snares containing "hackled bands," or broad threads of silk issuing from the cribellum. The fibers may be arranged in a regular manner or spun irregularly. The ribbon-like feature is only apparent under high magnification, and to the unaided eye the fibers appear like single threads. Actually, the supporting part, or warp, is made up of two or more longitudinal threads, and the sheet-like portion, or woof, is supported by the warp.

FAMILY HYPOCHILIDAE

This is a small family, with only a single genus represented in our region. There have been four species described, each named in honor of a distinguished student of spiders: Tord Thorell, Alexander Petrunkevitch, Pierre Bonnet, and Willis Gertsch. The general appearance and color is the same in all four. While the chelicerae are diaxial they are less completely so than in most labidognath spiders (fig. 54), and as is the case with tarantulas, the venom glands lie entirely within the chelicerae. The calamistrum consists of two rows of setae occupying the proximal third of metatarsus IV. These spiders make irregular meshed webs, sometimes shaped like a lamp shade, on the under surface of overhanging ledges, usually near a stream or in caves.

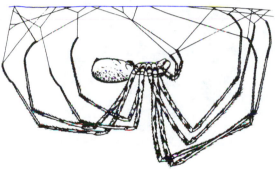

Figure 181. *Hypochilus thorellii,* lateral view of female in web.

Figure 181. *Hypochilus thorellii* Marx

The general color is grayish yellow with irregular dark purplish brown blotches. The cephalothorax is quite flat, higher in the eye region with a triad of eyes on either side and a pair of anterior median eyes by themselves between. The legs are extremely long, especially the anterior pair. Length of female about 14 to 15.5 mm; of male 10 to 11 mm.

Mountainous areas of North Carolina, Kentucky, Tennessee, Georgia and Alabama.

Hypochilus gertschi Hoffman
Length of female 15 to 20 mm; of male 14 mm.
The mountainous areas of Virginia and West Virginia.

Hypochilus petrunkevitchi Gertsch
Length of female 10 mm; of male about the same.
California.

Hypochilus bonneti Gertsch
Length of female 12.4 mm; of male 8 mm.
The mountainous areas of Colorado.

FAMILY FILISTATIDAE

The members of this small family build snares under stones, etc., and are often found in or around houses that are not kept up, the threads being quite conspicuous, especially around cracks and crevices. The spider constructs a tubular retreat in which it hides (fig. 182).

There are three genera. The common one is *Filistata,* with six species, all similar in appearance. The females are uniformly brown to black, often with irregular dusky blotches on the carapace. The males are uniformly yellowish tan, with very long pedipalps. The eyes are close together on a raised prominence (fig. 79). The tracheal spiracle is removed from the spinnerets a distance about one-third that to the epigastric furrow (fig. 80). The calamistrum is short (fig. 81). The spiders occur in warmer areas of our region and there are indications that females may live for eight years.

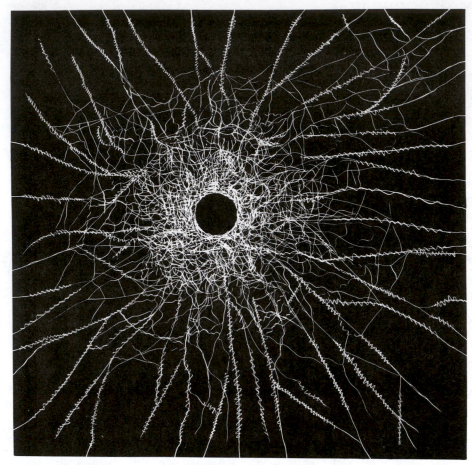

Figure 182. *Filistata hibernalis* web.

Figure 183. *Filistata hibernalis,* female.

Figure 183. *Filistata hibernalis* Hentz

Length of female 13 to 19 mm; of male 9 to 10 mm.

      Southern States west to eastern Texas.

Filistata arizonica Chamberlin & Ivie
Somewhat darker than *hibernalis*. Length of female 15 mm; of male 7 mm.

      Arizona and California.

Filistata utahana Chamberlin & Ivie
This species is darker than either of the other two above.

      Length of female 13 mm; of male 9.6 mm.

      Utah and New Mexico west to California.

FAMILY OECOBIIDAE

The members of this small family have three tarsal claws, and have both the carapace and sternum wider than long. The cribellum may be rudimentary and the calamistrum even absent in the males. The spiders live outside under stones, but are often found in houses, especially in cold climates. There are two genera.

There is a striking similarity between the Oecobiidae and the ecribellate family Urocteidae of the Old World. Some American workers prefer to consider them both as subfamilies within the same family, in which case the name Oecobiidae, having priority, should be used. Aside from the very fundamental difference as evidenced by the presence or absence of calamistrum and cribellum, only minor external differences can be found. The similarity extends not only to general appearance and external morphology, but to internal structure as well. This would appear to indicate a parallelism, or convergent evolution, a condition existing with a number of cribellate and ecribellate families and analogous to the similarity between certain marsupial and placental groups among the Mammalia.

1a **Tibia I about six or seven times as long as wide. Calamistrum from one-half to two-thirds the length of metatarsus IV** . . . . . .
. . . . *Oecobius* (**7 species**)

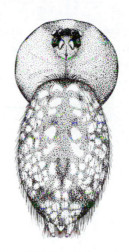

Figure 184. *Oecobius annulipes.*

Figure 184. *Oecobius annulipes* Lucas

The carapace is pale yellow with a marginal black line and dark spots and darker areas in front of and behind the eyes. The abdomen is white or light brown with many dark spots and the venter is pale. The calamistrum has a double row of bristles and extends about half the length of metatarsus IV. Length of female 2.5 to 2.9 mm; of male 2 to 2.6 mm.

This spider makes a small flat web on windowsills, and over cracks on the walls of buildings.

New England south to Florida and west to the Pacific.

Oecobius cellariorum (Dugès)
The carapace is virtually unmarked. The anterior median eyes are five-sixths the diameter of the posterior laterals, or smaller. Length of female 2.9 mm; of male 2.2 mm.

Maryland south to North Carolina and west to Kansas and Arizona.

Oecobius putus O.P.-Cambridge
As in *cellariorum* the carapace is virtually unmarked, but the anterior median eyes are much larger than the posterior laterals. Length of female 3.2 mm; of male 2.6 mm.

Texas west to California.

1b Tibia I only about four times as long as wide. Calamistrum as long as metatarsus IV . . . . *Platoecobius* (1 species)

Figure 185. Carapace of *Platoecobius*.

Figure 185. *Platoecobius floridanus* (Banks)

The sides of the carapace are sinuate. The color is tan, but with darker marginal bands. The abdomen is darker along the sides of the dorsum, and the venter is almost white. The posterior median eyes are nearly half the diameter of the posterior laterals. Length of female 2.4 mm; of male 2 mm.

South Carolina to Florida.

FAMILY ULOBORIDAE

These spiders spin geometric orb webs, or sectors of orbs, similar to those of the Araneidae and related families. They are unique among spiders in lacking poison glands.

1a Cephalothorax oval, longer than wide, with the two eye rows about the same length and eyes subequal (fig. 186). Tarsus IV more than half as long as metatarsus IV (fig. 187) . . . . *Uloborus* (9 species)

Figure 186. *Uloborus* carapace.

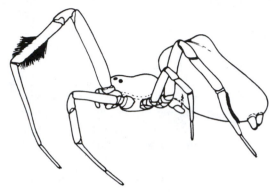

Figure 187. *Uloborus glomsus* from the side.

GENUS ULOBORUS

The members of this genus spin a complete orb, almost always in a horizontal plane, and about 100 to 150 mm in diameter. The spider does not build a retreat but strengthens the web with a stabilimentum or sheeted hub.

Figure 187. *Uloborus glomosus* (Walckenaer) Feather-legged Spider.

In females the tibiae I bear a bunch of hairs at their distal ends. Leg I is 4.5 times as long as the carapace. The abdomen of the female is high and bears a pair of humps at its summit about one-third of its length from the notched anterior end. The carapace of the female is about one and a third times as long as broad. The posterior row of eyes is so strongly recurved that a line along the front edges of the laterals does not touch the medians. The general color is grayish brown. Length of female 2.8 to 5 mm; of male 2.3 to 3.2 mm.

Southern Canada south to Florida and west to Nebraska and Texas.

Uloborus diversus Marx

Similar to *glomosus,* but with the carapace of the female nearly as broad as long, and leg I only 4.2 times the length of the carapace. Also, metatarsus I has a dark ring at the apex, and the brush on tibia I is weak or absent altogether. Length of female 3.2 to 4.7 mm; of male 2.1 to 2.7 mm.

Utah and west Texas west to Oregon and California.

Uloborus oweni Chamberlin

With the posterior row of eyes at most moderately recurved so that a line along the front edges of the laterals touches, or cuts through a portion of, the medians. The posterior medians are separated by two or more diameters. Leg I is five times the length of the carapace and tibia I lacks the brush. Length of female 3.7 to 5.7 mm; of male 3.4 to 4.2 mm.

Colorado, New Mexico and Arizona.

Uloborus octonarius Muma

Similar to *oweni* in the curvature of the posterior row of eyes, and in lacking the brush on tibia I. However, the posterior median eyes are separated by less than two diameters. Leg I is 6.5 times as long as the carapace, which in the female is nearly as broad as long. Length of female 3 to 5.8 mm; of male 2.8 to 3.3 mm.

Maryland west to Missouri and south to Tennessee and Alabama.

1b **Cephalothorax angular, as wide in the middle as it is long; with the posterior eye row much longer than the anterior, the posterior eyes much larger than the anterior, and the posterior medians farther lateral than the anterior laterals (fig. 188). Tarsus IV less than half the length of metatarsus IV. (fig. 189)** . . . . . .
. . . . *Hyptiotes* **(4 species)**

Figure 188. *Hyptiotes,* eye region from above.

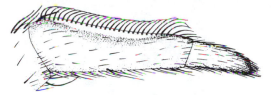

Figure 189. *Hyptiotes,* leg IV.

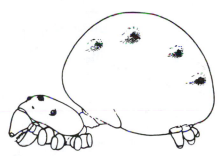

Figure 190. *Hyptiotes cavatus* from the side.

Figure 190. *Hyptiotes cavatus* (Hentz) Triangle Spider.

The abdomen of the female is broadly elliptical, provided with a double row of rounded bumps. The male has a narrower abdomen and the bumps are less prominent, and the palpal organ is very large and projecting. It resembles in color the brownish gray ends of the branches among which it lives, so that it is often not seen until one locates the web. The latter is easily recognized because of its triangular shape (fig. 191) being a sector of about 45° of an orb in a vertical plane. The line upon which the four radii converge is attached to a twig and the spider usually stands, dorsum down, quite close to the twig so that it resembles a bud. Length of female 2.3 to 4 mm; of male 2 to 2.6 mm.

New England to Florida and west to Texas and Missouri.

Figure 191. *Hyptiotes* snare.

Hyptiotes gertschi Chamberlin & Ivie
Similar to *cavatus* but the females have the second pair of tubercles on the abdomen quite prominent and occasionally greatly enlarged. Length of female 3.1 to 4.4 mm; of male 2.1 to 2.8 mm.

New England and adjacent Canada, New York, the Great Lakes States, the Rockies and Pacific Coast States north to Alaska.

FAMILY DICTYNIDAE

This is the largest family of cribellate spiders with nine genera in our region. Most of the members are small in size, and superficially resemble the members of the family Theridiidae in general appearance. They possess relatively very large poison glands. They build irregular snares in foliage in the tops of weeds and twigs, and underneath stones and dead leaves on the ground.

1a **Tarsi each with one long trichobothrium. Anterior median eyes quite small (to absent altogether). Clypeus low, little if any more than the diameter of an anterior median eye, and head region not especially elevated (fig. 192). Endites (a) short and not convergent (fig. 193) . . . .**
. . . . *Lathys* **(9 species)**

Figure 192. *Lathys,* face and chelicerae.

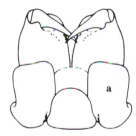

Figure 193. *Lathys,* underside of mouthparts.

Figure 194. *Lathys foxii.*

Figure 194. *Lathys foxii* (Marx)

The carapace is pale yellow, the abdomen dirty white with gray markings arranged as chevrons on the posterior two-thirds. The anterior median eyes are very small. Length of female 2.2 mm; of male 2 mm.

This spider makes its webs among dead leaves on the ground in wooded areas, from which it can be sifted.

New England and adjacent Canada south to Tennessee and west to Minnesota.

Figure 195. *Lathys maculina,* face and chelicerae.

Figure 195. *Lathys maculina* Gertsch

This species looks very much like *foxii* in pattern, but it has only six eyes, the anterior medians being absent. It is very common in the leaf litter of hardwood forests. Length of female 1.3 to 2.0 mm; of male 1.4 mm.

New York south to Florida and west to Texas.

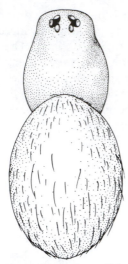

Figure 196. *Lathys pallida.*

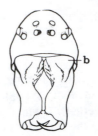

Figure 197. *Dictyna sublata* male, face and chelicerae.

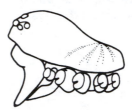

Figure 198. *Dictyna sublata* male, cephalothorax from side.

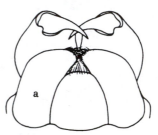

Figure 199. *Dictyna,* underside of mouthparts.

Figure 196. *Lathys pallida* (Marx)

This is another six eyed species, but without the gray markings shown by the preceding two. The general color is pale yellow. Length of female 1.2 to 1.6 mm; of male about the same.

New England and adjacent Canada south to Maryland and west to Wisconsin.

1b **Tarsi without trichobothria. Anterior median eyes hardly, if at all, smaller than the others. Clypeus considerably exceeding the diameter of the anterior eyes (fig. 197). Head region elevated, especially in males (fig. 198). Endites (a) long, and at least moderately convergent over labium (fig. 199)** 2

2a **Retromargin of cheliceral fang furrow with one tooth** . . . .
. . . . *Dictyna* **(120 species)**

GENUS DICTYNA

This is a very large genus. In the male the chelicerae are long, concave in front, and bowed outward near the middle of their length, and provided near the clypeal margin with a more or less well developed mastidion (b) (See figs. 197 and 198). The calamistrum occupies the middle half to two-thirds of the length of metatarsus IV (fig. 200). The color patterns are similar in related species yet variable within the same species. The male is usually much darker than the female.

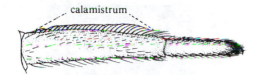

Figure 200. *Dictyna*, leg IV showing calamistrum.

Figure 202. *Dictyna volucripes.*

Figure 201. *Dictyna sublata.*

Figure 201. *Dictyna sublata* (Hentz)

In the female the cephalothorax is yellow in front and brown on the thoracic part. The abdomen usually has a broad median yellow band and is brown to the sides of this. The legs are yellow. In the male there is less difference between the cephalic and thoracic parts and the legs and abdomen are darker than in the female. Length of female 2.3 to 3.7 mm; of male 2 to 2.5 mm.

Eastern States and adjacent Canada west to Texas and the Dakotas, building its webs on the upper surface of large leaves.

Figure 202. *Dictyna volucripes* Keyserling

The cephalothorax is dark brown with two or three faint lines formed by rows of light hairs extending back from the eyes. The legs are a lighter brown than the carapace and are covered with gray hairs. On the abdomen is an elongate dark median band in front and irregular pairs of spots behind. The lighter areas are yellowish but the whole abdomen is covered with gray hairs. Length of female 2.5 to 4 mm; of male 2.7 to 3 mm.

Eastern States and adjacent Canada west to Colorado, building its snares on the ends of grass and weeds and sometimes on walls and fences.

Dictyna coloradensis Chamberlin

Very similar to *volucripes,* but is somewhat more brightly colored, and darker. It also averages somewhat larger, the females being about 3.8 mm; the males 3.2 mm.

New England and adjacent Canada west to Washington and Oregon and south through the Rocky Mountain and western Plains States.

Figure 203. *Dictyna foliacea.* (Hentz)

Figure 203. *Dictyna foliacea* (Hentz)

Similar to *sublata* but smaller. The carapace is light brown, lighter on the head. The abdomen is yellow in the middle and brown to gray toward the sides. There is much variation in the width of the central band. Length of female 2 to 2.7 mm; of male 1.7 to 2.1 mm.

The webs are built in grass, bushes and trees. Eastern States and adjacent Canada west to the Dakotas and Texas.

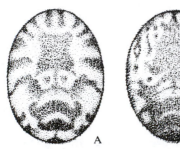

Figure 204. *Dictyna annulipes.*

Figure 204. *Dictyna annulipes* (Blackwall)

Grayer than *volucripes* with the middle markings on the abdomen wider in front and more broken behind, but the extent of the dark areas varies considerably, and two varieties are figured here. Length of female 2.9 to 4.4 mm; of male 2.4 to 3.8 mm.

Makes its snares in grass and weeds and on fences and walls. (fig. 205).

New England south to Virginia and across the northern States to Oregon and Washington, as well as all Canada to Alaska.

Figure 205. *Dictyna annulipes* snare.

Dictyna hentzi Kaston

This species closely resembles *annulipes*. It averages slightly smaller in size, however. Length of female 2.1 to 2.8 mm; of male 2 to 2.3 mm.

New England and adjacent Canada west to Minnesota and south to Mississippi and Texas.

Dictyna reticulata Gertsch & Ivie

The carapace is yellowish brown, darker on the sides of the cephalic portion. The abdomen is usually unmarked, varying from milky white to gray, but occasionally reticulated with dusky marks. Length of female 2.5 to 4 mm; of male slightly less.

South Dakota south to Texas and west to the Pacific Coast States. It has been reported as an important predator of cotton insects in California.

Dictyna bellans Chamberlin

The carapace is bright orange-red, but somewhat duller on the thoracic part. The abdomen is most often whitish with a brownish pattern consisting of a median dark mark widened behind into a transverse spot and followed by a double row of dark spots connected by narrow transverse bars. Length of female 2 to 4 mm; of male 2 mm.

Ohio south to Alabama, west to Wisconsin and South Dakota, and south to Utah and Arizona.

Dictyna calcarata Banks

The carapace is uniformly orange or dark reddish brown with black radiating lines on the sides. The abdomen is yellow or white with black markings. There is a narrow triangular mark from the front to the middle of the dorsum; two irregular bands from center to apex; and a specking of small dots on the sides. Length of female averaging about 3.5 mm; of male about 3 mm.

Western States from Montana south to Oklahoma and Texas, west to the Pacific Coast States. It frequently makes its unkempt webs on buildings.

2b Retromargin of cheliceral fang furrow with two, or three, teeth *Mallos* (8 species)

GENUS MALLOS

The members of this genus are all western and in general resemble the members of the genus *Dictyna*.

Mallos niveus O.P.-Cambridge

The carapace is brown with white lateral bands on the thoracic part. The abdomen is gray, with a dark brown band on the anterior half. Length of female 1.7 to 3.5 mm; of male slightly less.

Arizona and New Mexico northwest to Idaho and Washington.

Mallos pallidus (Banks)

The carapace is brown above with the white bands as in *niveus,* and the abdomen is gray. Length of female 2.25 to 5 mm; of male 3 mm.

Texas northwest to Montana and west to the Pacific Coast States.

Mallos trivittatus (Banks)

The carapace is uniformly brown lacking the white bands found in the preceding species. The abdomen is gray with a dark brown median stripe. Length of female 4.8 to 8 mm; of male 4 to 5.7 mm.

New Mexico north to Wyoming and west to California.

FAMILY AMAUROBIIDAE

These spiders closely resemble in general appearance members of the ecribellate family Agelenidae. By some workers they are included as belonging to the Dictynidae, and by others, especially the British, as having the name Ciniflonidae. There are 11 genera.

1a Trichobothria inconspicuous, short, hardly extending above the other hairs, and not increasing in length distally (fig. 206). Calamistrum virtually as long as metatarsus IV. (Anterior median eyes subequal to the posterior medians or only slightly smaller. Anterior lateral eyes the largest) . . . .
. . . . *Titanoeca* (4 species)

Figure 206. *Titanoeca,* leg IV, showing short tarsal trichobothria and long calamistrum.

GENUS TITANOECA

The members of this genus resemble one another in having the carapace orange, and the abdominal dorsum gray to black.

Figure 207A. *Titanoeca brunnea* dorsal view of female.

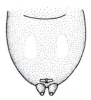

Figure 207 B. Ventral aspect of abdomen.

Figure 207. *Titanoeca brunnea* Emerton

This is the only species of the genus which shows a pair of white spots on the venter (fig. 207B). Specimens have been found under loose stones and dead leaves in areas drier than those preferred by *Callobius bennetti*. Length of female 4.5 to 5.5 mm; of male 4 to 5 mm.

New England and adjacent Canada south to Georgia and west to Arkansas and Missouri.

Titanoeca americana Emerton
Similar in appearance to *brunnea,* and often confused with it. However, it lacks the white spots on the venter. In the female the metatarsi I and II show only two distal spines at the end when viewed from below. Length of female 4.5 to 7.5 mm; of male 3.5 to 7 mm. The habits are similar to those of *brunnea.*

New England and adjacent Canada south to Virginia and west to New Mexico.

Titanoeca nigrella (Chamberlin)
Similar to *brunnea,* though lacking the two white spots on the venter. In the female the metatarsi I and II show three distal spines at the end when viewed from below. Length of female 5 to 8 mm; of male 4.5 to 7 mm.

Western half of the United States and Canada from the Mississippi to the Pacific.

Titanoeca silvicola Chamberlin & Ivie
Similar to *nigrella* but differing in showing four, rather than three, distal spines under metatarsi I and II. Length of female 4.5 to 7 mm; of male 4.5 to 5.5 mm.

Alaska south to Arizona and New Mexico.

1b Trichobothria conspicuous on tarsi and metatarsi, with their lengths increasing distally on each segment (fig. 208). The length of the calamistrum at most seven-tenths the length of the metatarsus IV . . 2

2a Anterior median eyes almost twice as large as the posterior medians. The calamistrum occupies the middle six to seven-tenths of the metatarsus IV (fig. 209). (Retromargin of cheliceral fang furrow with 2 teeth). . . . .
. . . . *Ixeuticus* (1 species)

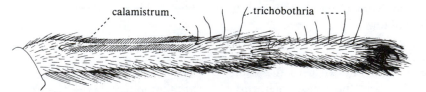

Figure 208. *Callobius,* leg IV, showing the trichobothria and the *apparently* double-rowed calamistrum.

Figure 209. *Ixeuticus martius,* leg IV showing calamistrum.

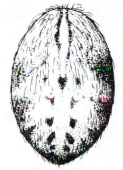

Figure 210. *Ixeuticus martius.*

Figure 210. *Ixeuticus martius* (Simon)

The carapace is dark brown, almost black along the sides. The abdomen is yellowish brown to gray, with a dark longitudinal mark at the front, and dark chevrons on the posterior half. Length of female 7.5 to 14 mm; of male 5 to 11 mm.

California.

2b Anterior median eyes smaller, at most only slightly larger than the posterior medians. Calamistrum not, or hardly, more than half the length of metatarsus IV . . . 3

3a Calamistrum *apparently* double; posterior to the calamistrum itself is an area of other bristles on the dorsal surface of the metatarsus, and the anterior edge of this batch of bristles resembles another row of calamistral bristles (fig. 208). Anterior median eyes subequal to, or slightly larger than the posterior medians. Calamistrum about half the length of

metatarsus IV. (Retromargin of cheliceral fang furrow with three or four teeth.) . . .
. . . . *Callobius* **(24 species) and** *Amaurobius* **(25 species)**

GENUS CALLOBIUS AND GENUS AMAUROBIUS

The members of these two genera look much alike and must be differentiated by characters of the genitalia. Most have the carapace brown, yellow or orange, and the abdominal dorsum gray, either unmarked or with light spots or chevrons usually (fig. 211). The calamistrum is about half the length of metatarsus IV, and is vestigial in males. The spiders are found chiefly under stones, leaf litter, and bark, and in rock fissures and stumps. Figure 212 shows the tangled web.

Figure 211. *Callobius bennetti.*

Figure 211. *Callobius bennetti* (Blackwall)

The carapace is brown, darkest on the cephalic part, with the legs lighter brown. The abdomen has a pattern of gray and white as illustrated. Length of female 5 to 12 mm; of male 5 to 9 mm.

Newfoundland south to Georgia and west to Manitoba and Tennessee.

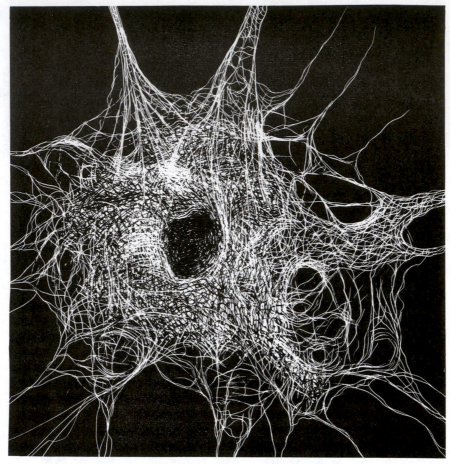

Figure 212. *Callobius* snare.

Callobius severus (Simon)
The carapace is orange-red. The abdomen is gray, either unmarked, or in some cases with one or two chevrons. Length of female 9 to 21 mm; of male 8 to 15 mm.
    British Columbia south to California.

Callobius pictus (Simon)
Similar to the preceding species. Length of female 8 to 18 mm; of male 8 to 10 mm.
    Alaska south to California.

Callobius nomeus (Chamberlin)
Similar to the preceding species. Length of female 9 to 11 mm; of male 8 to 9 mm.
    Northern New England northwest to Alaska.

Callobius nevadensis (Simon)
Similar to the preceding species, with the abdomen usually unmarked. Length of female 8 to 20 mm; of male 7 to 12 mm.
    Utah and Montana west to the Pacific Coast States.

Amaurobius ferox (Walckenaer)
This species looks very much like *Callobius bennetti* shown in Figure 211. Length of female 8.5 to 14 mm; of male 8 to 12.5 mm

New England and adjacent Canada south to Virginia and west to Illinois and Michigan.

Amaurobius borealis Emerton
The carapace is yellowish brown. The abdomen is reddish brown to gray with a lighter median band on the anterior half and chevrons on the posterior half. Length of female 4.3 to 6 mm; of male 3.5 to 5 mm.

New England and New York west to Minnesota, and in Canada all the way from Newfoundland west to British Columbia.

2b No such second row of bristles so that calamistrum is definitely uniseriate in appearance. Length of calamistrum about one-third that of metatarsus IV. Anterior median eyes smaller than posterior medians . . . .
. . . . *Callioplus* (8 species)

Figure 213. *Callioplus tibialis,* dorsal view of female.

Figure 213. *Callioplus tibialis* (Emerton)

The carapace and legs are brown, the abdomen brown with a lighter median band having crenate edges. The calamistrum is somewhat less than one-third the length of metatarsus IV. Length of female 5 to 9 mm; of male 5 to 6.5 mm.

From the mountainous areas of New England and eastern Canada to North Carolina.

Callioplus euoplus Bishop & Crosby
The carapace is orange, and the abdomen gray with lighter spots and chevrons. Length of female 4 to 7 mm; of male 3.5 to 5 mm.

Northern New England west to Minnesota, and in Canada from Newfoundland west to the Northwest Territory.

Section Ecribellatae

All of the remaining spiders in this work, lacking the cribellum and calamistrum, belong here. They produce several kinds of silk, some viscid and some non-viscid.

FAMILY SCYTODIDAE

Spitting Spiders

In this family the poison glands are enormously developed, occupying a very large part of the cephalothorax. It has been shown that each gland has two parts: a smaller part in front producing venom, and a larger posterior portion secreting a mucilaginous substance. The spider procures its prey by spitting at it, thus fastening the latter down with the gummy secretion. There is a single genus with seven species in our area.

Figure 214. *Scytodes thoracica,* female.

Figure 214. *Scytodes thoracica* (Latreille)

The general color is yellow with black markings. Those on the carapace are arranged so that the design vaguely resembles a lyre. Length of female 4 to 5.5 mm; of male 3.5 to 4 mm.

This species does not build a snare. It is often found in houses walking about slowly in shaded corners, dark closets and similar places. Some females have been known to live for as long as three years.

Eastern States and adjacent Canada to Indiana.

Figure 215. *Scytodes perfecta,* carapace.

Figure 215. *Scytodes perfecta* Banks

The carapace has a pattern of dark brown markings against a yellow background. The abdomen is provided with a series of dark spots against a white background, somewhat resembling the situation in *thoracica*. Length of female 6.5 to 8.8 mm; of male 5.8 to 6.5 mm.

Texas to California.

Figure 216. *Scytodes fusca,* carapace.

Figure 216. *Scytodes fusca* Walckenaer

The carapace is mostly dark brown, though in the male there is greater contrast with a median light stripe. The abdomen appears in most specimens to be more or less evenly brownish purple with some light areas mixed in. Some lack the light areas. Length of female 5.5 to 6 mm; of male 4.5 mm.

Florida.

FAMILY LOXOSCELIDAE

The members of this small family construct small irregular webs under logs, stones, etc., but also get indoors where they hide in dark corners, in trunks, under stored clothing, etc.

From records on laboratory kept specimens it is known that they may live for several years.

In recent years the members of our single genus *Loxosceles* have received much publicity because of the numerous cases of envenomation (see page 25). The venom of *L. reclusa* seems to be much stronger than that of *deserta* and *arizonica*. Only a few cases of bites by the latter two species have been recorded. Our species resemble one another closely in general appearance, nearly all showing to some degree a mark on the cephalic portion that resembles a violin. They are best distinguished by genitalia characters, but the relative length of leg I to the carapace can be used to some extent. There are six species in our area,[1] and in addition the large dark South American *L. laeta* (Nicolet) has been imported at several localities. In 1969 large numbers were found in several suburban areas of Los Angeles, California. This is the largest species of the genus, and reportedly possesses the most virulent venom.

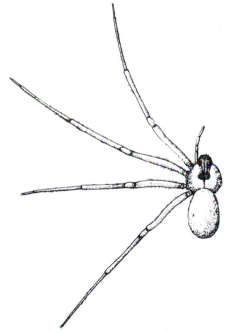

Figure 217. *Loxosceles reclusa*.

Figure 217. *Loxosceles reclusa* Gertsch & Mulaik. Brown recluse or violin spider.

The general color is yellowish brown, with a characteristic darker mark on the carapace, broader at the front and narrowed behind so that it resembles a violin (hence the common name). In females leg I is almost 4.5 times as long as the carapace, femur I being slightly longer. In males leg I is about 5.5 times the length of the carapace, the femur alone being one and two-thirds as long. Length of female averages 9 mm; of male averages 8 mm.

Ohio south to Georgia and west to Nebraska and Texas.

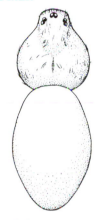

Figure 218. *Loxosceles deserta*.

Figure 218. *Loxosceles deserta* (Gertsch)

For many years this species had been mistakenly identified as *unicolor*. The carapace is light orange yellow with sparse pubescence. There is on some specimens a faint indication of the violin mark as in *reclusa*. The legs are light in color near the body but darken gradually distally. The abdomen is pale brownish gray. In females leg I is 5.5 times the length of the carapace, its femur being 1.6 times as long. In males leg I is 7.5 times the length of the carapace, its femur alone twice as long. Length of female 7.5 mm; of male 6 mm.

Arizona, Utah and Nevada.

1. There are a number of additional species which may soon be described.

Loxosceles arizonica Gertsch & Mulaik
Similar in general appearance to *unicolor*. In females leg I is four times as long as the carapace, its femur being slightly longer. In males leg I is almost six times the length of the carapace, its femur being one and a half times as long. Length of female averages 8 mm; of male slightly less.

Arizona.

Loxosceles rufescens (Dufour)
Similar in general appearance to the above three species. In females leg I is five times as long as the carapace, its femur 1.3 times as long. In males leg I nearly seven times as long as the carapace and its femur is nearly twice as long. Length of female 7.5 mm; of male 7 mm.

Apparently imported from Europe, but found in a number of localities in New York to Michigan and through the southeastern States.

FAMILY DIGUETIDAE

This is a small family of spiders restricted to our southwest. Our one genus *Diguetia* includes six species.

GENUS DIGUETIA

The webs consist of a maze of threads under cacti and other desert bushes about one to two feet off the ground. At the center of the web is a vertical tubular retreat closed at the top. The egg sacs are placed within this tube.

Figure 219. *Diguetia canities,* female.

Figure 219. *Diguetia canities* (McCook)

Metatarsus I is shorter than IV, and femur I of the male is less than twice as long as the carapace. The carapace is orange with white pubescence. The legs are yellow, with some orange brown on the patellae and the distal ends of the other segments, more deeply pigmented in the male than the female. The abdomen is tan with a white bordered folium on the dorsum and covered with white pubescence. Length of female 8 to 9.5 mm; of male 5.6 to 6.2 mm.

Oklahoma and Texas west to California.

Diguetia albolineata O.P.-Cambridge
In general appearance similar to *canities,* but smaller, and with proportionately longer legs. Metatarsus I is equal in length to IV, and femur I of the male is twice as long as the carapace. The posterior end of the abdomen is produced into a tail-like tubercle. Length of female 6.3 mm; of male 5.1 mm.

Texas west to California.

FAMILY PLECTREURIDAE

The members of this small family are known only from the west and there are two genera.

The spiders build a silken tube as a retreat, with an entrance ringed with silk. They are found under stones, and debris, in fractures of decomposed granite, in small holes along the edges of streams and roads, in the crevices of stone walls, etc. They hide in the daytime, but may come to the entrance at night.

1a **Femur I robust, curved, typically shorter than the carapace, and without dorsal spines, except rarely at base** . . . .
. . . . *Plectreurys* (10 species)[2]

Figure 220. *Plectreurys tristis,* female.

1b **Femur I more slender, essentially straight, much longer than the carapace, and with a dorsal series of spines along its entire length** . . . . *Kibramoa* (7 species)

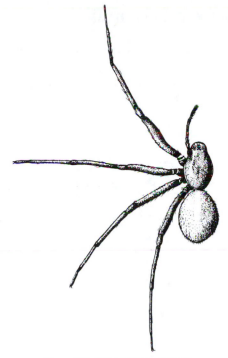

Figure 221. *Kibramoa suprenans.*

Figure 220. *Plectreurys tristis* Simon

The posterior eye row is slightly recurved. The cephalothorax and legs are dark mahogany brown, unmarked. The carapace is granulate, without hairs, and the abdomen is yellowish gray with gray pubescence. Length of female 12.5 to 17 mm; of male 11.5 to 17 mm.

Idaho, Utah, Nevada, Arizona and California.

Plectreurys castanea Simon

Similar in appearance to *tristis,* but smaller. The posterior row of eyes is weakly procurved or straight. Length of female 7.6 mm; of male the same.

California.

Figure 221. *Kibramoa suprenans* Chamberlin

The carapace and legs are brown, smooth and shiny. Metatarsus I of male has four pairs of ventral spines; of female has 15 or more. The abdomen is light gray, with a pale median stripe. Length of female 8 to 9.5 mm; of male 7.7 mm.

Arizona and California.

Kibramoa guapa Gertsch

Similar to *suprenans* but larger, with the carapace orange-brown, and the legs bright orange or red. Metatarsus I of the female has 11 ventral spines; of the male has a single pair

2. With the possibility of many more species soon to be described.

apical in position. Length of female 12.4 mm; of male 9 mm.

California.

FAMILY OONOPIDAE

1a **Abdomen covered with soft cuticle, without scuta. Eye group transverse with the laterals separated by the posterior medians, which are the largest** . . . .
. . . . *Orchestina* **(4 species)**

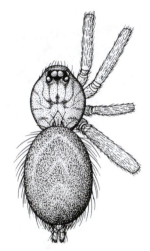

Figure 222. *Orchestina saltitans.*

Figure 222. *Orchestina saltitans* Banks

The cephalothorax is convex with the highest point at the middle. The carapace is yellow with irregular purple markings and a black line along the lateral margins, and black around the eyes. The legs are yellowish orange, and the abdomen purplish with many long fine hairs. Femur IV is considerably thicker than the other femora. The spinnerets arise from a low peduncle. Length about 1 mm, with little difference between the sexes.

These spiders have been recorded as having been taken in sweeping grass and bushes outdoors. However, they seem to be most commonly taken inside of buildings. Even though this is one of the smallest known species, the sharp-eyed can see specimens moving about slowly over walls or tables, or hanging from lampshades, in buildings. They can be recognized easily by their marked jumping ability, an ability correlated with the great development of the hind femora. To induce them to jump one needs only to touch them gently.

New England south to Georgia and west to Missouri.

Orchestina moaba Chamberlin & Ivie
Similar to *saltitans,* but somewhat larger and lighter in color. Length of female 1.7 mm; of male 1.3 mm.

Utah, Arizona and California.

1b **Abdomen with a dorsal scutum, divided in the female into longitudinal parts. Eye group elliptical with the anterior laterals the largest and virtually contiguous** . . . . .
. . . . *Scaphiella* **(2 species)**

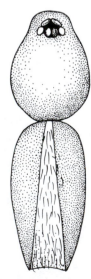

Figure 223. *Scaphiella hespera,* Female.

Figure 224. *Scaphiella hespera,* abdomen of male.

Figures 223 and 224. *Scaphiella hespera* Chamberlin

The general color is orange over the entire body. In the male the dorsal scutum appears complete, though thicker laterally. In the female the scutum is split longitudinally to expose a narrow less sclerotized area which is wider behind and clothed with hairs. The sternum is truncate behind and widely separates the coxae IV. Length of female 1.7 to 2 mm; of male 1.5 to 1.9 mm. Found under detritus on the ground.

Texas west to California in arid and semi-arid areas.

FAMILY DYSDERIDAE

This is a small family, with a single genus *Dysdera,* and a single widely distributed species.

Figure 225. *Dysdera crocata.*

Figure 225. *Dysdera crocata* C. L. Koch

The cephalothorax and legs are reddish orange and the abdomen dirty white, with few or no hairs over the body. The chelicerae are long and project forward considerably. The six eyes are arranged in a transverse oval. The coxae of legs I and II are cylindrical, longer and thinner than those of legs III and IV. The labium is longer than wide, and the endites are practically parallel. Length of female 11 to 15 mm; of male 9 to 10 mm.

The spiders live under stones, sometimes under bark and in moss; for the most part seeking dark and humid surroundings. They construct a flattened, oval retreat of tough silk, and the eggs are laid in a very light transparent cocoon. The spiderlings live with their mother for a while. No snare is built, the spiders hunting their prey from the retreat.

New England south to Georgia and west to California.

FAMILY SEGESTRIIDAE

This is a small family, and by some workers is considered just a subfamily of the Dysderidae.

There is only one genus common enough to be included here, *Ariadna,* with four species in our region.

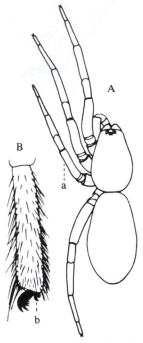

Figure 226. *Ariadna bicolor.* A, dorsal view; B, lateral aspect of tarsus 1 showing 3 claws.

Figure 226. *Ariadna bicolor* (Hentz)

The cephalothorax is brown, relatively long and narrow. The abdomen is a long oval, mostly purplish brown above and below, paler at the sides. Length of female 6.1 to 15 mm; of male 5.4 to 10.6 mm.

The spider builds a tubular retreat in cracks of trees, rocks, etc., and under bark and stones (fig. 227).

New England south to Florida and west along central and southern States to Colorado and California.

FAMILY PHOLCIDAE

The anterior median eyes are smallest (fig. 228) (and in one group entirely lacking), while the other eyes are arranged in two triads. The legs are extremely long and thin and the tarsi flexible. The snares are either sheet-like or irregular, the spider hanging in an inverted position beneath the web. They are built in cellars and other dark locations. The eggs when laid are surrounded by only a few fibers and held by the female in her chelicerae (fig. 229). There are six genera.

Figure 227. *Ariadna bicolor* retreat.

Figure 228. *Pholcus,* carapace from above.

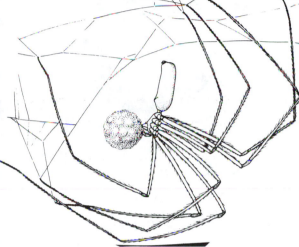

Figure 229. *Pholcus* female holding her bag of eggs.

1a   With six eyes, in two triads, not elevated on a prominence (fig. 230). Abdomen globose . . . .
. . . . *Spermophora* (1 species)

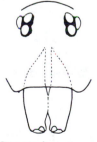

Figure 230. *Spermophora,* eyes and chelicerae.

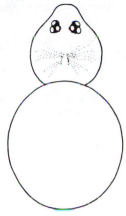

Figure 231. *Spermophora meridionalis,* female.

Figure 231. *Spermophora meridionalis* Hentz. Short-bodied Cellar Spider.

The entire body is pale yellow except for a pair of light gray spots on the carapace. Length of female 2 mm; of male 1.6 mm.

Eastern half of the United States west to Missouri.

1b   With eight eyes elevated on a prominence. Abdomen elongated, or oval, or triangular seen from the side 2

2a   Abdomen elongate, more than twice as long as wide, and more than twice as long as the cephalothorax. Femur I five or more times as long as the carapace. . . . . .
. . . . *Pholcus* (2 species)

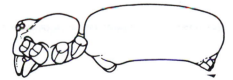

Figure 232. *Pholcus phalangioides.*

Figure 232. *Pholcus phalangioides* (Fuesslin) Long-bodied cellar spider.

The color is pale yellow except for a gray mark in the center of the carapace. Length of female 7 to 8 mm; of male 6 mm.

This is the commonest cellar spider throughout the United States.

2b Abdomen almost as wide as long and less than twice as long as cephalothorax. Femur I less than five times as long as carapace **3**

3a Posterior eye row recurved, the eyes subequal. Anterior lateral eyes slightly larger than the posterior medians. Femur IV as long as I, or slightly longer. (Abdomen in most cases appears triangular from the side, highest behind at spinnerets, but not prolonged much beyond them.) . . . . *Physocyclus* **(4 species)**

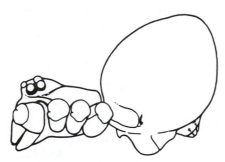

Figure 233. *Physocyclus globosus.*

Figure 233. *Physocyclus globosus* (Taczanowski)

The abdomen is pale gray with an indication of darker spots on either side of the midline. The cephalothorax is dirty yellow, the legs light yellow, with darker rings at the distal ends of femora and both ends of tibiae. Length of female 4.7 mm; of male 3.7 mm.

A house spider of the warmer regions from Florida west to California.

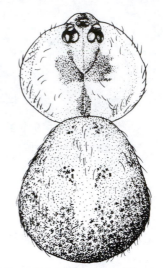

Figure 234. *Physocyclus californicus.*

Figure 234. *Physocyclus californicus* Chamberlin & Gertsch

The carapace is yellow with a brown Y-shaped mark at the center. The abdomen is darker, brown to black behind with paired spots as illustrated. Length of female 5.3 to 7 mm; of male 4.8 to 6 mm.

California.

3b Posterior eye row slightly procurved, the medians slightly the larger, and larger than the anterior laterals. Femur I longer than IV . . . . *Psilochorus* **(14 species)**

GENUS PSILOCHORUS

The carapace is pale except for a forked dark mark along the cervical and dorsal grooves. The abdomen is light, with four or five pairs of dark blotches on the dorsum. Each of the blotches is made up of a group of small pigmented spots. In specimens long preserved these may look black, but in fresh material they may appear tinged with green, blue or purple. In most cases the abdomen appears oval from the side, with about two-thirds of its length extending posterior to the spinnerets.

Figure 235. *Psilochorus pullulus.*

Figure 235. *Psilochorus pullulus* (Hentz)

The abdomen is yellow with four pairs of greenish purple spots above. The carapace is yellow except for a forked black mark along the dorsal and cervical grooves. Length of female 3.2 mm; of male 2.9 mm.

Maryland south to Georgia and west to Texas.

Psilochorus utahensis Chamberlin
Length of female 3.5 to 4.4 mm; of male 3.3 to 3.5 mm.

Utah, Arizona and California in arid and semi-arid areas.

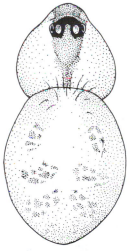
Figure 236. *Psilochorus californicus.*

Figure 236. *Psilochorus californicus* Chamberlin

Length of female 2.7 to 3.2 mm; of male 2.2 to 2.8 mm.

California.

FAMILY THERIDIIDAE

Comb-footed Spiders

There are 24 genera in our region. The members of this very large family build irregular snares (fig. 237) from the threads of which they suspend themselves in an inverted position while awaiting their prey. Viscid silk is flung over the prey by the activity of the hind legs. The tarsi of these latter are provided on the ventral surface with a row of 6 to 10 slightly curved serrated

Figure 237. *Theridion* snare.

bristles composing a comb (fig. 238). Although this comb is the outstanding characteristic of the family, these bristles may be inconspicuous and often difficult for the novice to recognize and the comb is absent from some.

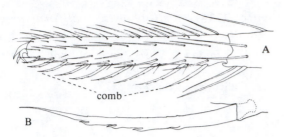

Figure 238. *Theridion*. A, Tarsus IV showing comb; B, Single bristle from comb.

Only after the prey is enswathed and relatively quiet does the spider approach close enough to bite it. Most commonly the victim is dragged to that part of the web in which the spider usually takes up its station. It has been observed on numerous occasions that very small theridiids can render *hors de combat* with their viscid silk insects many times their own size.

The beginner may have difficulty separating the members of this family from those of the Linyphiidae, and should note the following which may be of help: Besides lacking the comb on tarsus IV, the members of the latter family usually have thinner legs, which are usually provided with spines. The chelicerae are more robust and their margins always provided with teeth; the labium has its edge much thickened (i.e., strongly rebordered); the endites are usually parallel; and the height of the clypeus *usually* much exceeds that of the median ocular area. In the Theridiidae the tarsal comb is present; the legs may be moderately to very long, but are either completely devoid of spines, or at least have none on tibiae and metatarsi; the chelicerae are in most cases not very robust and with the margins commonly unarmed with teeth; the labium is not rebordered (except in *Conopistha* and *Rhomphaea* where it is fused to the sternum); the endites are slightly converging; and the height of the clypeus *generally* does not exceed that of the median ocular area. The two families differ also in the structure of the genitalia.

1a **Males with dorsal (a) and epigastric (b) scuta (fig. 239). Females with parts of epigastric scutum above pedicel and over each lung cover. (Sternum broadly truncate between coxae IV. Body length less than 2 mm. Carapace finely granulate and tarsal comb may be difficult to see)** . . . . .
. . . . *Pholcomma* (3 species)

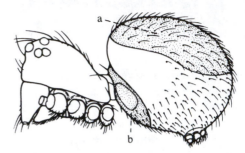

Figure 239. *Pholcomma*, male, from side.

Figure 240. *Pholcomma hirsutum*, female.

Figure 240. *Pholcomma hirsutum* Emerton

The abdomen is whitish to pale gray. In the male there is a brown dorsal scutum which covers almost the entire dorsum; in the female this is lacking. The cephalothorax and legs are brown with indistinct gray markings. The chelicerae are provided with 3 or 4 minute den-

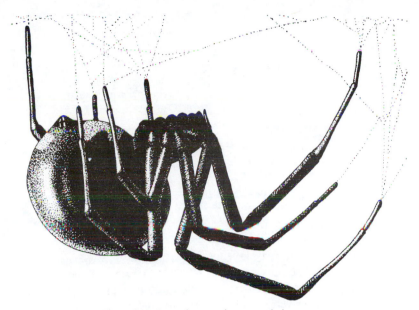

Figure 241. *Latrodectus* as it appears in its snare.

ticles on each margin of the fang furrow. Length of female 1.3 to 1.6 mm; of male the same.

Collected from forest floor litter, New England south to Florida and west to Missouri and Iowa.

1b   Scuta absent **2**

2a   Lateral eyes of each side at least a diameter of one apart. Abdomen globose, shining, usually black (with red, or red and white spots, which are seldom lacking) . . . .
. . . . *Latrodectus* **(5 species)**

GENUS LATRODECTUS
Widow Spiders

In North America north of Mexico there are five species. Of these *L. geomstricus* C.L. Koch, the "brown widow," is cosmotropical in its distribution, whereas the "red-legged widow," *L. bishopi* Kaston, appears to be limited to southern Florida.

The "black widow" for many years was thought to be a single species but is now known to be represented by three species, which are very similar in appearance and habit, and each is quite variable in itself. The body of the female is shiny, and usually black or almost so. Most commonly there are red spots and sometimes white ones as well on the globose abdomen. The most characteristic spot appears on the venter and is somewhat in the shape of an hourglass. Most specimens also have a red spot just above the anal tubercle. The male has the abdomen much narrower and lower, and usually with more red and white marks. Spiderlings are much lighter in color, with very little or no black. After each molt a larger area is covered with black pigment so that females half to two-thirds grown may resemble in color and pattern full grown males, which generally mature after fewer molts than do females.

These are the most notorious of all spiders. It is true that the venom is highly virulent (see page 24) but the spider is quite timid, and has no instinct to bite humans. Even when disturbed in its web it attempts to escape rather than to attack. Its web is an irregular mesh usually built quite close to the ground with a strong-walled funnel-shaped retreat in a dark spot sheltered from the weather. The webs are found commonly in spaces under stones or logs, or holes in dirt embankments, and occasionally in barns, rural privies, and other outbuildings.

As in other species the female may kill and eat the male after mating, but in spite of popular belief the male is not always eaten. Ordinarily, if the females are well fed the males manage to get away safely and have been known to mate again with other females. During the course of a summer a female may lay several masses of eggs (fig. 242). She may produce egg sacs the following year, and some females may live for more than three years.

Figure 242. *Latrodectus* egg sacs: left L. mactans; right *L. variolus* and *L. hesperus.*

Figure 243. *Latrodectus mactans* female.

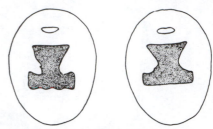

Figure 244. *Latrodectus mactans,* ventral aspect of abdomen showing two variations in hourglass markings.

Figure 245. *Latrodectus mactans,* male side view.

Figures 243, 244, and 245. *Latrodectus mactans* (Fabricius)

The female usually shows a full hourglass with its posterior half more generally a rounded rectangle than a triangle, and wider than the anterior half. There is usually a row of red spots along the mid-dorsum.

The egg sacs are usually nearly spherical, about 11 or 12 mm in diameter, and generally show a conspicuous nipple at the top (fig. 242). The sacs usually show a grayish tinge even when freshly made. This species is the smallest of the three black widows, females for the most part averaging 8 to 10 mm in length, and males 3.2 to 4 mm.

Southern New England to Florida west to eastern Oklahoma, Texas and Kansas. More common in the southern part of the range.

Figure 246. *Latrodectus variolus* female.

Figure 247. *Latrodectus variolus,* ventral aspect of abdomen showing two variations in hourglass markings.

Figure 248. *Latrodectus variolus,* male side view.

Figures 246, 247, and 248. *Latrodectus variolus* (Walckenaer)

In this species the hourglass is usually present in two parts not joined at the middle. Also, in addition to the row of red spots along the mid-dorsum there are diagonal white bands laterally on the dorsum. The egg sacs when fresh are at most only very pale gray, often yellow to tan, and more pear shaped, most often spread at the top, about 13 or 14 mm in length, and about 10 to 12 mm in diameter (fig. 242).

The female is slightly larger than *mactans,* averaging 9 to 11 mm, while the male is the largest of the three species, averaging 5.5 to 6 mm.

New England and adjacent Canada south to Florida, and west to eastern Texas, Oklahoma and Kansas. More common in the northern part of the range.

Figure 249. *Latrodectus hesperus,* female.

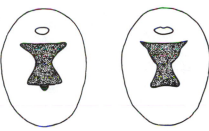

Figure 250. *Latrodectus hesperus,* ventral aspect of abdomen showing two variations in hourglass markings.

Figure 251. *Latrodectus hesperus,* male side view.

Figures 249, 250, and 251. *Latrodectus hesperus* Chamberlin & Ivie

In this species the hourglass mark is generally complete, with both halves triangular, and the anterior triangle longer and broader than the posterior. There is seldom any red mark on the dorsum additional to the one just above the anal tubercle, and in many specimens even this may be absent. The egg sac is similar in size and shape to that of *L. variolus,* but never has the grayish tinge, being just creamy yellow to tan (fig. 242). The female is the largest of the three species, averaging 10.5 to 13 mm, but the males are intermediate, averaging 3.5 to 4.5 mm.

Western portions of Texas, Oklahoma and Kansas north to the adjacent Canadian provinces and west to the Pacific Coast States.

2b   Lateral eyes of each side less than a radius apart **3**

3a   Posterior median eyes three or more times the diameter of one of them apart . . . . *Spintharus* **(1 species)**

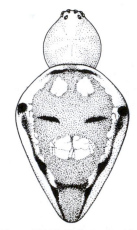

Figure 252. *Spintharus flavidus.*

Figure 252. *Spintharus flavidus* Hentz

The cephalothorax and legs are yellow to orange. The abdomen is marked with black, red and yellow stripes and spots. Length of female 4 to 4.5 mm; of male 2.8 mm.

Taken from the underside of leaves in low bushes. Eastern states to Oklahoma and Texas.

3b   Posterior median eyes rarely more than twice their diameter apart **4**

4a   Carapace with a broad and deep transverse furrow on the thoracic part. Abdomen prolonged above and beyond the spinnerets so that the spinnerets are at least as far from the distal end as from the pedicel. (Tarsal comb may be difficult to see.). **5**

4b   Carapace without such a transverse furrow. Abdomen not greatly prolonged beyond spinnerets **6**

5a   Posterior row of eyes procurved with the posterior medians much farther from each other than from the posterior laterals. Metatarsus I shorter than tibia I. Clypeus almost horizontal. Abdomen long and vermiform (fig. 253C). (Head of male not provided with horns) . . . . *Rhomphaea* **(2 species)**

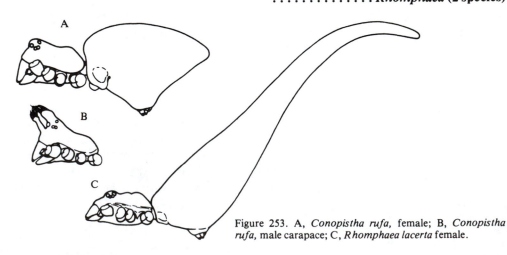

Figure 253. A, *Conopistha rufa,* female; B, *Conopistha rufa,* male carapace; C, *Rhomphaea lacerta* female.

Figure 253C. *Rhomphaea lacerta* (Walckenaer)

This species is also known under the name *fictilium* (Hentz). The general color is light yellow with silver markings and with three darker bands on the carapace. Length of female 5 to 10.5 mm; of male 4 to 7 mm. The length is variable in accordance with the degree of extension of that part of the abdomen beyond the spinnerets, which exceeds the rest of the body in length. The legs are quite long, the first longer than the body.

Found in bushes and grass.

New England south to Florida, west to Missouri and Texas; also in Pacific Coast States.

5b **Posterior row of eyes in an almost straight line and eyes almost equidistant. Metatarsus I not shorter than tibia I. Clypeus more nearly vertical. Abdomen not vermiform. (Head of male with two horns)** . . . . *Conopistha* (14 species)

Figure 253A and B. *Conopistha rufa* (Walckenaer)

This species is also known under the name *Argyrodes trigona* (Hentz). The general color is brownish yellow, sometimes with metallic reflections. Length of female 3.7 to 4.2 mm; of male 2.5 to 3.3 mm. The abdomen is drawn out beyond the spinnerets and though the back is usually straight it has been said that the spider can turn down the tip. In the female the ocular area is raised and separated by a notch from the clypeus. In the male a long horn is produced on either side of this notch and the abdomen is less angular than in the female.

Though they can build their own webs these spiders have most often been taken from the webs of other spiders. The egg sacs are very characteristic in appearance, being vase or bottle-shaped, about 6 mm long and suspended in the web by a thread about twice that length

(fig. 254). The sacs are white when first made, later changing to brown, and like the spiders themselves may easily be mistaken for scales from pine buds.

Taken from pine and spruce trees.

Eastern states and adjacent Canada.

Figure 254. Egg sac of *Conopistha rufa*.

6a **Abdomen elongated, subtriangular, and more or less pointed behind. (Leg IV longer than I. Colulus setae absent.)** . . . . *Euryopis* (20 species)

Figure 255. *Euryopis limbata*, female.

Figure 255. *Euryopis limbata* (Walckenaer)

This species is also known under the name *funebris* (Hentz). The cephalothorax is yellowish gray and very wide. The abdomen widens nearly to the middle of its length, then narrows to a point above the spinnerets. Most of the dorsum is covered by a black area pointed behind and bordered by silver on the posterior half. Length of female 3 to 4 mm; of male 3 mm.

These spiders are found under leaves and moss on the ground.

New England and adjacent Canada south to Georgia and west to Wisconsin and Louisiana.

6b **Abdomen rounded behind, not elongated or subtriangular.** 7

7a **Colulus lacking; its presence not even indicated by setae** 8

7b **Colulus present, or its presence indicated by two setae (fig. 269)** 12

8a **Abdomen (in female at least) wider than long, with a hump or tubercle on each side at about the middle of its length. (Posterior median eyes a little larger than posterior laterals. Anterior eye row procurved.).** . . . . *Theridula* (3 species)

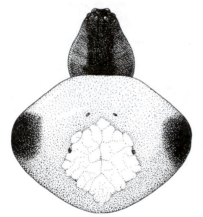

Figure 256. *Theridula emertoni* Levi.

Figure 256. *Theridula emertoni* Levi

In the female the carapace is yellowish gray on the sides with a broad black band in the middle, and the legs are yellow. The abdomen is greenish gray with a white spot in the middle, a large black spot around the spinnerets and one

on each side where the abdomen is drawn out into a hump. In the male the carapace and legs are orange and the median black band is much more distinct. The abdomen is much narrower than in the female. Length of female 2.3 to 2.8 mm; of male 2 to 2.3 mm.

Taken from bushes and hemlock trees.

New England and southern Canada south to Tennessee and west to Wisconsin.

Theridula opulenta (Walckenaer)
Very similar to *emertoni* in size and pattern.

New York south to Florida and west to Texas and also in Utah and Oregon.

8b **Abdomen without lateral humps and not wider than long** 9

9a **Abdomen of female higher than long in most cases** 10

9b **Abdomen of female not higher than long.** 11

10a **Abdomen usually with a pattern of spots on the anterior half, and with a mid-dorsal white stripe on the posterior half extending to the anal tubercle (fig. 257). Epigynum with a cone-shaped knob so that from the side it appears beaked. The males are tiny, in length at most one-third that of the females, and with only a single pedipalp which is disproportionately large (fig. 257)** . . . . *Tidarren* (2 species)

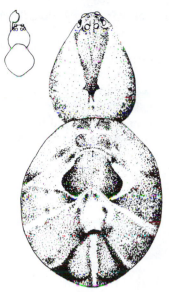

Figure 257. *Tidarren sisyphoides,* male and female to same scale.

Figure 257. *Tidarren sisyphoides* (Walckenaer)

The color is very variable, from a brownish tan to a grayish black. The legs are spotted and banded. The female constructs a retreat in which she hides and places her egg sacs. Length of female 5.8 to 8.6 mm; of male 1.2 to 1.4 mm.

North Carolina south to Florida and west to California.

10b **Abdomen without the mid-dorsal white stripe. Some specimens have a posterior dorsal tubercle. Epigynum not beaked. Males not tiny, their length at least half that of the females, and with both pedipalps of the more usual proportions .**
. . . . *Achaearanea* **(14 species)**

GENUS ACHAEARANEA

The common members of this genus are still frequently referred to by the name *Theridion,* a large genus from which they have been removed.

Figure 258. *Achaearanea tepidariorum,* abdomen.

Figure 258. *Achaearanea tepidariorum* (C.L. Koch) House Spider; Domestic Spider.

The carapace is yellowish brown and the abdomen dirty white to brown with indistinct gray chevrons on the posterior half. The legs of the male are orange; of the female yellow with dusky annuli at the ends of the segments. Some individuals have a conspicuous spot in the center of the dorsum. Length of female 5 to 6 mm; of male 3.8 to 4.7 mm.

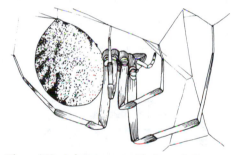

Figure 259. *Achaearanea tepidariorum* in its snare.

This extremely common spider is found most often in barns and houses where it makes its webs (see fig. 237) in the corners of rooms and in the angles of windows. It has been collected outside under stones and boards, on bridges and fences. It usually stands in its web (fig. 259) where a part of the latter is more closely woven than the rest, but does not construct a "tent." Adults can be found at all seasons and some individuals may live for a year or more after becoming mature. The egg sacs are familiar objects in the webs. They are brownish, ovoid or suborbicular, sometimes pear shaped, of 6 to 9 mm in diameter, with a tough papery cover.

Throughout the United States and also Canada.

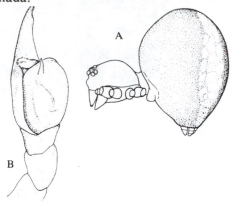

Figure 260. *Achaearanea globosus.* A, Female; B, Male palp.

Figure 260. *Achaearanea globosus* (Hentz)

The carapace and legs are orange. The abdomen of the female is higher than it is long, is broadest at the highest part and more or less pointed behind. There is a large black spot in the middle of the dorsum, and another in the middle of the venter. Length of female 1.6 to 2 mm; of male 1.2 to 1.8 mm.

Found in leaf litter, at edges of logs or in holes of tree stumps, etc. Its egg sacs are peculiar in being lozenge shaped, pointed at both ends; they are cream-colored and placed freely in the web.

New England and adjacent Canada south to Florida and west to Texas and Minnesota.

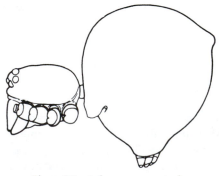

Figure 261. *Achaearanea rupicola.*

Figure 261. *Achaearanea rupicola* (Emerton)

The general color is dirty gray with darker gray and brownish markings. Easily recognized by the small dorsoapical cone or tubercle in the middle of the dorsum. Length of female 1.8 to 2.9 mm; of male 1.4 to 2.2 mm.

Found under stones and boards in woods and around houses. The spider makes a retreat camouflaged with debris and inside of which it hides its discoidal white to brown egg sacs.

New England and adjacent Canada south to Alabama and west to the Mississippi River.

Achaearanea porteri (Banks)

The carapace is yellow to brown, darker along the margins and in the ocular area. The abdomen is irregularly spotted with black and gray. Some specimens resemble *rupicola* in having a tubercle on the abdomen. Length of female 2.2 to 4.9 mm; of male 1.6 to 2.8 mm.

New York south to Florida and west to Texas and Kansas.

11a **Total body length less than 2.5 mm. Legs short, patella plus tibia of leg I together less than the length of the carapace. Dorsum lacking a pattern, though there may be some black or gray spots . . . .**
. . . . *Thymoites* (12 species)

Figure 262. *Thymoites unimaculatum.*

Figure 262. *Thymoites unimaculatum* (Emerton)

The carapace and legs are orange, the eye region black. The abdomen is oval, white to orange with black around the spinnerets and on the mid-dorsum as illustrated. Length of female 1.9 to 2.1 mm; of male 1.8 to 2 mm.

The web is made in bushes and low plants near the ground and the egg sacs are white and spherical.

New England south to Florida and west to Texas and Minnesota.

11b Larger spiders. Legs longer, the patella plus tibia of leg I more than 1.5 times the length of the carapace. Dorsum with a pattern . . . . *Theridion* (69 species)

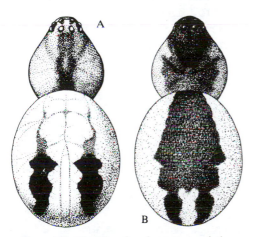

Figure 263. *Theridion frondeum,* two varieties.

Figure 263. *Theridion frondeum* Hentz

The ground color is white, light yellow, or pale greenish, with dark markings that vary greatly in their extent. The figures show a light and a dark variety. The carapace may have a narrow or broad central band, and the abdomen may have pairs of black spots or a broad median band. The legs are white with dark spots at the distal end of tibiae and metatarsi. The chelicerae of the male are each furnished with a

mastidion in front. Length of female 3 to 4.2 mm; of male 3 to 3.5 mm.

The web is built on the underside of leaves in shrubs, and also in tall grass. The egg sacs are spherical, about 3.7 to 4.4 mm in diameter and fastened in the web.

New England and adjacent Canada south to Alabama and west to North Dakota.

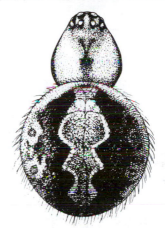

Figure 264. *Theridion murarium.*

Figure 264. *Theridion murarium* Emerton

This species is very similar to *differens* and *glaucescens*. Ths carapace is grayish yellow with a narrow black marginal stripe each side and a broader median stripe. The abdomen has a broad light median band with wavy margins and is gray to black on either side of the band. Length of female 2.8 to 4 mm; of male 2.7 to 3 mm.

The web is made in trees, in bushes and grass, and also under stones near the ground. The egg sacs are white to tan, spherical, about 3 to 4 mm in diameter and at first held by the hind legs then attached in the web and guarded.

Throughout the United States and southern Canada.

Figure 265. *Theridion differens.*

Figure 267. *Theridion lyricum,* female.

Figure 265. *Theridion differens* Emerton

Similar to *murarium* and *glaucescens,* but with a narrower median band on the dark brown to gray dorsum. Length of female 1.8 to 3.5 mm; of male 1.6 to 2.5 mm.

Throughout the United States and southern Canada.

Figure 267. *Theridion lyricum* Walckenaer

The carapace is brownish gray, somewhat lighter in a triangular area extending back from the posterior eyes to the dorsal groove. The legs are brownish gray to orange, banded in the male, and with the markings indistinct in the female. The abdomen has a median white band with a large black area either side of it at the anterior end. Length of female 3 to 3.5 mm; of male 2.1 to 2.8 mm.

New England south to Florida and west to Texas and Missouri.

Figure 266. *Theridion glaucescens.*

Figure 266. *Theridion glaucescens* Becker

This species is also known under the name *spirale* Emerton. It is similar to *murarium* and *differens.* The abdominal pattern is not as distinct as in the latter, but more distinct than in the former. In the females the central band is much wider than is the case in the other two species. Length of female 2.2 to 3 mm; of male 1.6 to 2.5 mm.

New England and adjacent Canada to Florida and west to Nebraska and Texas.

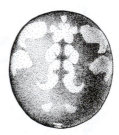

Figure 268. *Theridion alabamense,* female.

Figure 268. *Theridion alabamense* Gertsch & Archer

The carapace and legs are yellow suffused with gray. The abdomen is gray, marked in the center and on the sides with lighter areas which are smaller in the male. Length of female 1.9 to 3.7 mm; of male 1.8 to 2.7 mm.

New England south to Florida and west to Texas. Also in California.

12a **Colulus distinct and at least half as long as the setae extending from it** **14**

12b **Colulus indicated only by two setae (a) (fig. 269)** **13**

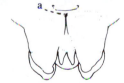

Figure 269. *Anelosimus* spinnerets.

13a **Chelicerae with several teeth on promargin of fang furrow. Clypeus not concave, and eye region not projecting** . . . .*Anelosimus* (2 species)

Figure 270. *Anelosimus textrix,* abdomen.

Figure 270. *Anelosimus textrix* (Walckenaer)

This species is also known under the name *studiosus* (Hentz). The carapace and legs are orange yellow with an indistinct median gray band extending forward from the dorsal groove, and forking there to send two branches to the eyes. The abdomen is gray along the sides and has a dark median band bordered in white. Length of female 3.2 to 4.7 mm; of male 2.1 to 2.3 mm.

This species has attracted attention because of the large masses of webbing made by members of a colony on bushes and trees. Col-ony formation is quite rare among spiders. Moreover, it displays another behavior trait known for very few spiders, i.e., the mother feeds the young spiderlings by regurgitation.

New England south to Florida and west to Texas.

13b **Chelicerae small, without teeth on promargin of fang furrow. Eye region in many specimens projecting forward over a clypeus which is quite concave (fig. 271)** . . . .*Dipoena* (17 species)

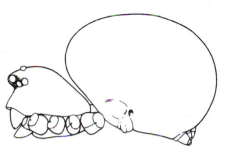

Figure 271. *Dipoena nigra.*

Figure 272. *Dipoena nigra,* female.

Figure 272. *Dipoena nigra* (Emerton)

The cephalothorax and legs are yellow to brown with gray markings. The abdomen is black. Length of female 2.8 to 4 mm; of male 1.5 to 2 mm.

Taken from trees and bushes and in sifting litter.

Throughout the United States and southern Canada.

14a **Sternum truncate behind and broadly produced between coxae IV. Carapace with numerous small crescent-shaped elevations each at one side of a puncture (fig. 273). (Body length less than 3 mm.)** . . . . *Crustulina* (2 species)

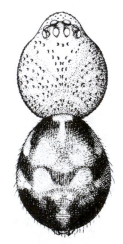

Figure 273. *Crustulina altera.*

Figure 273. *Crustulina altera* Gertsch and Archer

The cephalothorax is dark orange brown with the crescentic elevations darker and harder. The abdomen has a purplish brown color with light gray and white markings, the pattern varying in different individuals. Length of female 2.4 to 2.7 mm; of male 2.3 mm.

Found under logs and stones in pastures.

New England south to Florida and west to Louisiana, Kansas and Minnesota.

Crustulina sticta (O.P.-Cambridge)
This species has the carapace somewhat darker than in *altera*. Some specimens have the abdomen above pale yellowish to orange, without spots; but some have purplish brown spots on the anterolateral areas. All have a median broad dark purplish band on the venter. Length of female 2.3 to 2.7 mm; of male 2.2 mm.

New England and adjacent Canada west through the middle States to the Pacific Coast States and north to Alaska.

14b **Sternum pointed or evenly rounded (sometimes broadly) behind. Carapace without crescent-shaped elevations . . . 15**

15a **Retromargin of cheliceral fang furrow with two denticles. Abdomen unicolorous. Sternum narrowly rounded behind and not prolonged between coxae IV (fig. 274). Cymbium of male pedipalp elongate and much wider at the proximal than distal end (figs. 276, 277)** . . . . *Ctenium* (15 species)

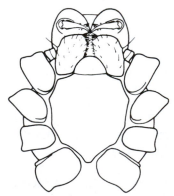

Figure 274. *Ctenium*, cephalothorax from below.

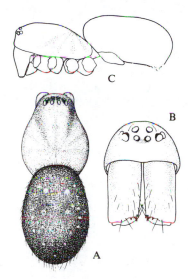

Figure 275. *Ctenium.* A, From above; B, From the side; C, Face and chelicerae.

GENUS CTENIUM

In this group (fig. 275) the chelicerae are robust with three promarginal teeth with two minute denticles on the retromargin. The species resemble one another closely, differentiation being based almost entirely upon genitalia characteristics. The carapace is yellowish brown to chestnut or reddish brown, and glabrous. The legs are brown or orange, somewhat larger on the femora. The abdomen is usually gray, but varies from grayish tan to black, and is covered with a sparse pubescence.

These spiders are mature at all seasons and live in moss, under stones and boards on the ground, and in forest floor litter from which they have been sifted.

Figure 276. *Ctenium riparius,* cymbium from above.

Figure 276. *Ctenium riparius* (Keyserling)

The cymbium of the male palp has two spines (a). Length of female 2.7 to 4.1 mm; of male 2.5 to 3.6 mm.

The commonest species.

New England and adjacent Canada south to North Carolina and Tennessee, and west to Wyoming.

Figure 277. *Ctenium banksi,* cymbium from above.

Figure 277. *Ctenium banksi* Kaston

The cymbium of the male palp has three apical spines (b). Length of female 3 to 4.2 mm; of male 2.8 to 3.3 mm.

New England and adjacent Canada south to Maryland and west to Wisconsin.

15b   Retromargin of cheliceral fang furrow either lacking teeth entirely, or with one tooth; or else the male chelicerae are much enlarged and with one to several retromarginal teeth (fig. 278). Abdomen

either with a definite pattern, or else spotted. Male with palpal cymbium oval. . . 16

16a Chelicerae of male quite robust (fig. 278). Chelicerae of female with a single tooth on the retromargin of the fang furrow . . . . *Enoplognatba* (9 species)

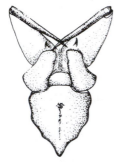

Figure 278. *Enoplognatha,* mouthparts and sternum of male.

Figure 279. *Enoplognatha marmorata.*

Figure 279. *Enoplognatha marmorata* (Hentz)

The cephalothorax and legs are an even yellowish brown and the abdomen is gray to white with markings that are variable. The sides of the folium are usually scalloped, and, in some specimens, silvery. The chelicerae are toothed on both margins in both sexes. The sternum is narrowly prolonged between coxae IV. Length of female 6 to 7 mm; of male 5 to 6 mm.

Found under rock ledges, stones and boards and occasionally under leaves, and in bushes near the ground.

Throughout the United States and southern Canada.

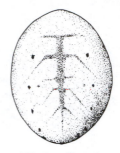

Figure 280. *Enoplognatha ovata.*

Figure 280. *Enoplognatha ovata* (Clerck)

This species appears in three varieties, each of which was described under a different name. It is unfortunate that priority belongs to the name of the least common variety. The carapace is yellow, with a black marginal stripe each side, and the legs are yellow with a black spot at the end of tibia I. The abdomen is pale with markings which are quite variable. The most common is the form *lineatum,* with several pairs of black spots as illustrated here. The form *redimitum* shows two red bands along the dorsum, converging behind to enclose an oval middle area. The form *ovata* has the entire middle area red. In the male the chelicerae are enormously developed, with long fangs and a long dentiform apophysis on the retromargin (fig. 278). Length of female 4.3 to 7 mm; of male 3.5 to 5.2 mm.

Northeastern States and adjacent Canada; Pacific Coast States and British Columbia.

16b Chelicerae of male not robust. Both sexes lacking teeth on the retromargin of the cheliceral fang furrow . . . . *Steatoda* (17 species)

GENUS STEATODA

Some of the species in our area were until recently known under the generic names *Teutana (triangulosa, grossa); Lithyphantes (albomaculatus);* and *Asagena (americana).* Most species show a light band across the front of the abdominal dorsum. The males have a more or less well developed stridulating organ composed of a series of transverse striations on the posterior part of the carapace which are scraped by a denticulated area on a sclerotized plate borne on the front of the abdomen as showing in Figure 281B.

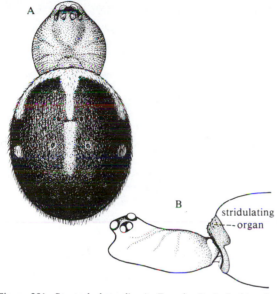

Figure 281. *Steatoda borealis*. A, Female; B, Left side of male to show stridulating organ on abdomen.

Figure 281. *Steatoda borealis* (Hentz)

The carapace is orange brown, covered with short stiff hairs. The sternum is darker and granulate while the legs are not quite as dark. The abdomen is a purplish brown to black with a yellow median line on the anterior half which joins the light line encircling the anterior half to form a "T." Length of female 6 to 7 mm; of male 4.7 to 6 mm.

Taken in sweeping low herbage, but usually from under bark and stones, in rock crevices, as well as on bridges and in corners of unused sheds, etc. The snare is characteristic (fig. 282).

New England and adjacent Canada southwest to Texas and west to the Rockies.

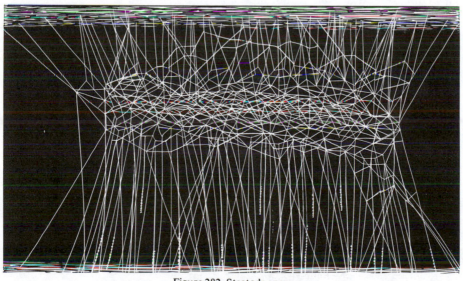

Figure 282. Steatoda snare.

Steatoda hespera Chamberlin and Ivie
Similar in appearance to *borealis,* but differing primarily in the structure of the genitalia. Length of female 4.2 to 7.5 mm; of male 3.6 to 5.4 mm.

The Rocky Mountain States west to the Pacific Coast States.

Figure 283. *Steatoda americana.*

Figure 283. *Steatoda americana* (Emerton)

The cephalothorax and legs are dark chestnut brown with the sternum granulate and somewhat darker than the carapace. The abdomen is dark purplish brown with a pair of white spots across the middle. Length of female 3.5 to 4.7 mm; of male 3.2 to 4.4 mm.

Found under stones and logs, and under loose bark and debris near the ground.

Eastern United States and adjacent Canada to the Pacific Northwest.

Figure 284. *Steatoda albomaculatus.*

Figure 284. *Steatoda albomaculata* (DeGeer)

The cephalothorax and basal leg segments are brown, the rest of the leg segments yellow with dark brown annuli at their ends. The abdomen is yellowish and with a very variable pattern of black markings. In the female there is a tooth on the promargin of the cheliceral fang furrow; in the male there is a large tooth on the retromargin and the chelicerae are powerfully developed. Length of female 4 to 8 mm; of male 4.3 to 6.8 mm.

The webs are built under stones and boards near the ground.

New England and adjacent Canada west to the Great Basin and the northern Rocky Mountain States, and the Pacific Coast States.

Figure 285. *Steatoda triangulosa.*

Figure 285. *Steatoda triangulosa* (Walckenaer)

The cephalothorax is orange brown and the legs yellow with darker annuli at the ends of the segments. The abdomen shows a pattern of purplish brown markings on a yellow ground. Length of female 3.6 to 5.9 mm; of male 3.5 to 4.7 mm.

Found under stones, culverts and bridges.
Throughout the United States.

Figure 286. *Steatoda grossa.*

Figure 286. *Steatoda grossa* (C.L. Koch)

Most specimens have the abdomen purplish brown with pale yellow markings as illustrated, but sometimes the light markings are inconspicuous or hardly at all present. These latter specimens have often been mistaken for black widows. However, the shape of the abdomen is more oval, and not as globose as in *Latrodectus*. Length of female 5.9 to 10.5 mm; of male 4.1 to 7.2 mm.

In the southern and western states this species has been reported making its snares commonly in dwellings, and has been reported preying upon the black widow spiders. It is one of the commonest "house spiders" of southern California, and may live for up to six years.

Coastal States of the Atlantic, Gulf, and Pacific regions.

Steatoda fulva (Keyserling)
The ground color of the carapace and legs is yellow to dark brown, with darker annuli on the legs. The abdomen is dark reddish brown, with white spots variable in arrangement. Usually there is a stripe around the anterior part of the dorsum, and on each side there is a series of lateral spots that may be joined to form a stripe. Length of female 3 to 5.9 mm; of male 2.4 to 5 mm.

These spiders are almost always found on the ground, with webs under stones and boards, and in holes in the ground.

Georgia and Florida to Texas, north to Nebraska, and west to Oregon and California.

FAMILY NESTICIDAE

This is a small family of widely distributed spiders which build their webs under stones, in dark places such as caves, etc. There are two genera in our area, each with a single species.

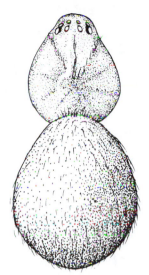

Figure 287. *Eidmanella pallida.*

Figure 287. *Eidmanella pallida* (Emerton)

The carapace and legs are shiny orange to brown. The abdomen is lightly pubescent, purplish gray, with very faint light purplish markings. Length of female 3.5 to 4 mm; of male 3 to 3.5 mm.

New England south to Florida west to Kentucky and Tennessee; also from Arizona and California.

FAMILY LINYPHIIDAE

Line Weaving Spiders

The chelicerae are moderately powerful, lack a boss, and in some species the lateral surface is provided with a row of horizontal striae (a) making up a stridulating area (fig. 288). The margins of the fang furrow are oblique and armed with teeth. In the males of some species there is a toothlike cusp, or mastidion, on the front of the chelicera, near the clypeal margin (fig. 289). The legs are long and thin, and usually provided with fine spines. The labium is strongly rebordered, as in the Araneidae and

Tetragnathidae. The pedipalp of the male has a paracymbium.

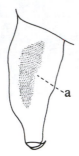

Figure 288. *Meioneta,* chelicera showing stridulating striae.

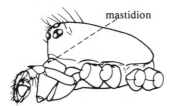

Figure 289. *Frontinella* male showing mastidion.

Some of the smaller members of this family may be confused with those of the Micryphantidae, and it is often difficult to place with certainty in one or the other family some of the members of these two groups. Almost any character selected will be found to have exceptions.

Most species construct a snare of some sort, usually with a platform, or a dome, as well as with an irregular portion. No retreat, molting, or cocooning webs are built and the spider simply takes up a position on the under surface of the snare.

The beginner may have difficulty distinguishing some members of this family from the Theridiidae. The differences are given in the discussion of the latter family. There are 20 genera in our region.

1a   Promargin of cheliceral fang furrow with seven (or six) large teeth, of which the fourth from the medial side is the longest.

Anterior median eyes larger than posterior medians. Clypeus lower than the height of median ocular area (fig. 290) . . . . *Tapinopa* (2 species)

Figure 290. *Tapinopa* face and chelicerae.

Figure 291. *Tapinopa bilineata,* female.

Figure 291. *Tapinopa bilineata* Banks

The carapace is dark brown or gray on the sides with a median light line widest in front, and narrowing to a point behind. The abdomen has a double row of dark spots along its length and is darker behind. Length of female 5 mm; of male 4 to 5 mm.

In grass or leaves close to the ground.

New England and adjacent Canada south to Georgia and west to Wisconsin.

1b Margins of cheliceral fang furrow with not more than five or six teeth, which are not as prominent as in *Tapinopa*. Anterior median eyes not larger than the posterior medians. Clypsus higher than the height of the median ocular area . . . **2**

2a Chelicerae each with a set of three (or four) conspicuous spines on the prolateral face (fig. 292), the uppermost largest and lowermost smallest. Promargin of fang furrow with five (or six), and retromargin with four or five teeth . . . .
. . . . *Drapetisca* (2 species)

Figure 292. *Drapetisca* Face and chelicerae.

Figure 293. *Drapetisca alteranda.*

Figure 293. *Drapetisca alteranda* Chamberlin

The carapace is creamy gray, with a dark spot on each of the radial furrows and a dark margin each side. The abdomen is broadly oval, widest just beyond the middle of its length and pointed behind. It is creamy gray with several pairs of dark spots and a triangular pattern as figured. Length of female 4 to 4.5 mm; of male 3.2 to 3.8 mm.

Commonly found walking over the trunks of trees, the bark of which they resemble in color. The web is a sheet closely appressed to the tree trunk, hence difficult to see.

New England west to Wisconsin.

2b Chelicerae with no more than two spines on prolateral face of each, and usually less conspicuous than in *Drapetisca* **3**

3a Median ocular area with the sides practically parallel and the anterior median eyes about equal in size to the posterior medians **4**

3b Median ocular area noticeably wider behind, and the anterior median eyes smaller than the posterior medians. **7**

4a Stridulating file (a) on lateral face of chelicera (fig. 294) quite conspicuous in both sexes **6**

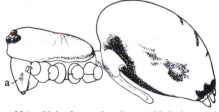

Figure 294. *Helophora* showing stridulating striae on chelicera.

4b Stridulating file inconspicuous and difficult to see **5**

5a Tarsus I about three-fourths as long as metatarsus I, which in males is swollen at its middle and slightly beyond so as to be noticeably thicker than metatarsus II. Promargin of cheliceral fang furrow with three, retromargin with two (or three) teeth. Male without a mastidion near clypeal margin and with a long spine on the patella of the palp . . . .
. . . . *Stemonyphantes* (1 species)

Figure 295. *Stemonyphantes blauveltae.*

Figure 295. *Stemonyphantes blauveltae* Gertsch

The colors are grayish yellow with broad median and narrow marginal black stripes on the carapace, and three rows of black spots on the abdomen. Length of female 5 to 6 mm; of male 4 mm.

Found under moss, in marshes, and under stones and debris.

New England and adjacent Canada south to Virginia and northwest to Washington.

5b Tarsus I about half to two-thirds as long as metatarsus I, which in males is not thicker than metatarsus II. Promargin of cheliceral fang furrow with four to six, retromargin with three to five teeth. Male with a mastidion near clypeal margin, and

with a short, black tooth on the palpal patella (see fig. 289). . . . .
. . . . *Frontinella* (2 species)

Figure 296. *Frontinella pyramitela.*

Figure 296. *Frontinella pyramitela* (Walckenaer). Bowl and Doily Spider.

This species is also known under the name *communis* (Hentz). The carapace is evenly brown. The abdomen is about as high above the spinnerets as in front and is marked above with a pattern as figured. Length of female 3 to 4 mm; of male 3 to 3.3 mm.

The characteristic bowl and doily webs (fig. 297) are made in pine woods, in bushes and tall grass.

Throughout the United States and Canada.

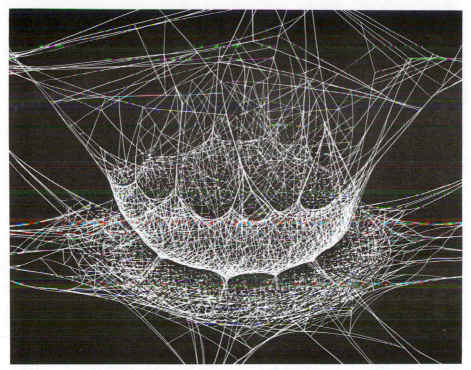

Figure 297. *Frontinella pyramitela* snare.

6a **Tibiae without lateral spines and very few spines elsewhere. Size under 3 mm. Palp of female without claw. Cymbium usually angulate** . . . . *Meioneta* **(22 species)**

Figure 298. *Meioneta micaria*, male.

Figure 298. *Meioneta micaria* (Emerton)

The carapace is yellowish brown in the center and dark brown to gray along the sides. The abdomen is variable commonly showing across the middle a broad white band pointing forward as illustrated. In others the white areas are larger so that there appears to be just a pair of black spots on each side of the middle. Length of female 1.9 mm; of male 1.5 to 1.8 mm. This species resembles *fabra* but the male lacks a mastidion and has the chelicerae (fig. 288) less robust than in that species.

New England south to Georgia and west to Illinois.

Figure 299. *Meioneta fabra*, male.

Figure 299. *Meioneta fabra* (Keyserling)

Similar to *micaria,* but the male has more robust chelicerae, with a mastidion (fig. 300). Length of female 2 to 2.5 mm; of male 1.8 to 2.4 mm.

New England to New York south to Alabama and Texas.

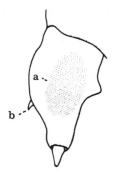

Figure 300. *Meioneta fabra,* chelicera of male showing stridulating striae (a) and also mastidion (b).

6b Tibiae with lateral spines and legs heavily spined in general. Size over 5 mm. Palp of female with claw. Cymbium not angulate . . . . *Pimoa* (6 species)

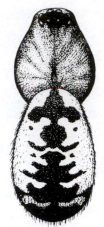

Figure 301. *Pimoa altioculata.*

Figure 301. *Pimoa altioculata* (Keyserling)

The carapace is yellow suffused with gray along the margins. The legs are yellow to brown with gray annuli, and the abdomen is pale dirty white with a median folium of black blotches. Length of female 6 to 7 mm; of male 5 to 6 mm.

Oregon north to British Columbia.

7a Stridulating file on lateral face of chelicera quite conspicuous in both sexes. Posterior median eyes separated by, at most, hardly more than a diameter. . . . 13

7b Stridulating file inconspicuous and difficult to see. Posterior median eyes separated by one and a half times a diameter, usually two or three times. . . . 8

8a Patella of male palp with a long, unpigmented dorsolateral spur (fig. 302). Epigynum with a short scape expanded distally, and separating two large openings (fig. 303) . . . . *Pityohyphantes* (13 species)

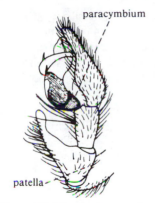

paracymbium

patella

Figure 302. *Pityohyphantes* palp.

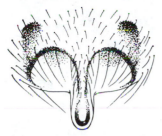

Figure 303. *Pityohyphantes* epigynum.

Figure 304. *Pityohyphantes costatus.*

Figure 304. *Pityohyphantes costatus* (Hentz). Hammock Spider.

The carapace is white or light yellow with thin black lines at the margins, and a broader forked one extending back from the posterior median

eyes. The abdomen has a conspicuous herring-bone dark brown central band along the midline. Length of female 5 to 7 mm; of male 4.5 to 6 mm.

The snares are built on fences, in garages and other outbuildings, on shrubs and the lower branches of trees.

New England south to North Carolina and west to the Pacific coast and across southern Canada.

8b Patella of male palp without a spur. Epigynum without a free scape **9**

9a Males with at least one mastidion near the clypeal margin. Epigynum with a single opening **10**

9b Males without a mastidion. Epigynum with a pair of openings. **12**

10a Posterior median eyes closer to each other than to the posterior laterals, and only scantily surrounded by pigment. Femora I and II without dorsal and lateral spines. (Male with a single mastidion on each chelicera.) ....
.... *Prolinyphia* **(3 species)**

Figure 305. *Prolinyphia marginata.*

Figure 305. *Prolinyphia marginata* (C.L. Koch). Filmy Dome Spider.

Some workers consider the correct name to be *Neriene radiata* (Walckenaer). The carapace is dark brown except for the lateral margins which are white. The abdomen is widest and highest behind and has a characteristic pattern of brown markings on a yellowish white background as figured. Length of female 4 to 6.5 mm; of male 3.4 to 5.3 mm.

The characteristic snares (fig. 306) of this very common species are seen in wooded areas, in underbrush and about rockpiles and stone walls.

New England south to Florida and west to the Pacific Coast States.

Figure 306. *Prolinyphia marginata* snare.

Figure 307. *Prolinyphia litigiosa.*

Figure 308. *Neriene clathrata.*

Figure 307. *Prolinyphia litigiosa* (Keyserling)

The carapace and legs are creamy white to pale green, and the former is without the strongly contrasting bands present in *marginata*. In males there might be a slightly pale orange tint. The sternum and venter are black. Length of female 5.2 to 8.5 mm; of male 5.1 to 6.8 mm.

Montana south to Utah and west to the Pacific Coast States and British Columbia, usually in mountainous areas.

10b **Posterior median eyes closer to the posterior laterals than to each other, and with a thick ring of black pigment around them. Femur I with dorsal and prolateral spines** **11**

11a **Femur I with two dorsal and one (or two) prolateral spines** ....
.... ***Neriene*** **(5 species)**

GENUS NERIENE

The species placed here have long been known as belonging in the genus *Linyphia,* from which they have recently been removed.

Figure 308. *Neriene clathrata* (Sundevall)

The carapace is evenly brown with the legs yellow to orange. The abdomen is lighter brown with a pattern of black chevron markings. The male has both an anterolateral and antero-medial mastidion on each chelicera. The female has three teeth on the promargin and 5 or 6 on the retromargin of the cheliceral fang furrow. Length of female 3 to 5.2 mm; of male 3.6 to 4.8 mm. Found in salt marshes, and in wooded areas close to the ground among forest litter and stones.

New England and adjacent Canada south to North Carolina and west to the Great Plains States.

Neriene variabilis (Banks)
This species is also known under the name *maculata* (Emerton). The carapace is orange yellow and the legs are somewhat lighter. The abdominal dorsum is pale in front with a few gray spots, which are larger and darker toward the posterior end. The male is somewhat darker, and has only an anteromedial masti-dion on each chelicera. The female has 4 or 5 teeth on each margin of the cheliceral fang fur-row. Length of female 3.4 to 5.4 mm; of male 3.8 to 4.7 mm.

New England south to Georgia and west to Minnesota.

11b **Femur I with only one dorsal and two prolateral spines. (Males with only one mastidion on each chelicera.).** ....
.... ***Microlinyphia*** **(4 species)**

Figure 309. *Microlinyphia mandibulata.*

Figure 309. *Microlinyphia mandibulata* (Emerton)

The carapace is evenly dark brown; the abdomen shows white spots as indicated, though often the spots are indistinct. The male has greatly elongated chelicerae, more than half as long as the carapace, and they are directed backward. The tibia of the male palp has only a dorsal spine. Length of female 4 to 5 mm; of male 3.3 to 4.6 mm.

The flat web is found in open meadows usually in grass, close to the ground.

There are two subspecies. *M. mandibulata mandibulata* is found from New England south to Georgia and west through the northern tier of States to Washington. *M. m. punctata* (Chamberlin and Ivie) occurs from British Columbia south to California and Arizona.

Microlinyphia pusilla (Sundevall)

Similar to *mandibulata,* but averaging a bit smaller. It is very variable in abdominal pattern, with the spots often indistinct. The tibia of the male palp has a retrolateral spine near the distal margin as well as a dorsal spine. Length of female 3 to 4.8 mm; of male 2.8 to 4 mm.

Colorado and Wyoming west to Idaho and Washington.

12a Abdomen without a folium; with a caudal tubercle (fig. 310) . . . .
. . . . *Florinda* (1 species)

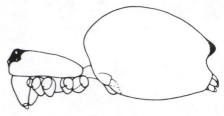

Figure 310. *Florinda coccinea.*

Figure 310. *Florinda coccinea* (Hentz)

The carapace is orange to red, except for black in the eye region. The abdomen is yellow, except for the black caudal tubercle. Length of female 3.5 mm; of male 3 mm.

The webs are built in grass.

Maryland west to Illinois and south to Florida and Texas.

12b Abdomen with a folium; without a caudal tubercle. . . . .
. . . . *Estrandia* (1 species)

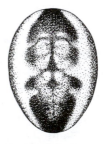

Figure 311. *Estrandia grandaeva.*

Figure 311. *Estrandia grandaeva* (Keyserling)

The carapace and legs are light brown. The abdomen has a white longitudinal stripe each side between which is a pattern of dark areas as figured. Length of female 2.4 to 2.8 mm; of male 2 to 2.5 mm.

The webs are built in evergreen trees.

New England and adjacent Canada south to North Carolina and west to the Rockies.

13a Tibia of male palp with a short lateral process bearing three stout hairs

(fig. 312). Epigynum with openings concealed, and with a scape that is about same width throughout . . . .
. . . . *Helophora* **(3 species)**

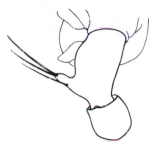

Figure 312. *Helophora*, showing bristles on tibia of male palp.

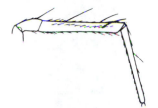

Figure 314. *Lepthyphantes*, patella, tibia and metatarsus.

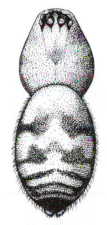

Figure 315. *Lepthyphantes zebra*.

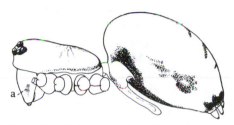

Figure 313. *Helophora insignis*, female from side.

Figure 313. *Helophora insignis* (Blackwall)

The carapace and legs are yellow; the abdomen is pale gray or white. Often the abdomen is unmarked but usually there are gray spots on the sides and near the posterior end. Length of female 3.2 to 4.4 mm; of male 2.8 to 3.7 mm.

The flat webs are built near the ground in grass and low bushes.

New England west to Utah and across southern Canada.

13b   Tibia of palp without the lateral process. Epigynum either without such a long scape, or with the openings not obscured and the scape more slender distally. . . . 14

14a   Metatarsi armed with one or more spines (fig. 314) . . . .
. . . . *Lepthyphantes* **(23 species)**

Figure 315. *Lepthyphantes zebra* (Emerton)

The carapace is yellow, darker along the margins. The abdomen is white to light gray with scattered silvery spots and several transverse dark bands. Often the bands are faint and indistinct. Length of female 2 to 2.2 mm; of male 1.8 to 2 mm.

The webs are constructed among the curled up dead leaves on the forest floor.

New England south to North Carolina and west to Washington. Also southern Canada.

Figure 316. *Lepthyphantes nebulosa.*

Figure 318. *Bathyphantes pallida.*

Figure 316. *Lepthyphantes nebulosa* (Sundevall)

The carapace is dull orange with a dark marginal line and a central dark stripe which forks in front toward the posterior median eyes. The abdomen is yellowish white to gray with several pairs of dark blotches as figured. Length of female 4 to 4.5 mm; of male 3.5 to 4 mm.

The snare is built close to the ground under stones and boards.

New England and adjacent Canada south to Georgia and west to Washington.

14b Metatarsi unarmed 15

15a Tibiae with lateral spines (fig. 317). Palp of female with tarsal claw . . . .
. . . . *Bathyphantes* (22 species)

Figure 318. *Bathyphantes pallida* (Banks)

The carapace is yellow to brown, and the abdomen gray or black with several pairs of transveres white spots as figured. Length of female 2.3 to 2.8 mm; of male 1.9 to 2.1 mm.

The snare is built quite close to the ground in wooded areas.

New England and adjacent Canada south to North Carolina and west to Illinois.

15b Tibiae without lateral spines. Palp of female without claw 16

16a Metatarsus IV with a trichobothrium (fig. 319). Male palp with femur expanded distally, and with cymbium not angulate. Spiracle in the usual place close to spinnerets . . . .
. . . . *Microneta* (15 species)

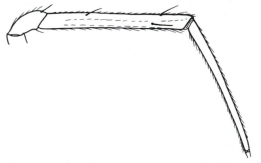

Figure 317. *Bathyphantes,* patella, tibia and metatarsus.

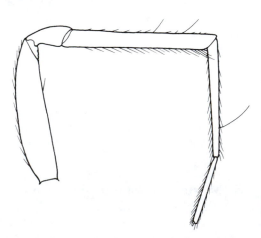

Figure 319. *Microneta,* leg IV showing lack of spines but presence of trichobothrium on metatarsus.

Figure 320. *Microneta viaria.*

Figure 321. *Tennesseellum* venter.

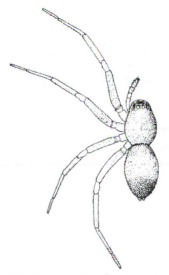

Figure 322. *Tennessesllum formicum,* female.

Figure 320. *Microneta viaria* (Blackwall)

The cephalothorax is brownish yellow with a fine black marginal line. The abdomen is gray to black without any definite markings, and sometimes with a slight iridescence. The legs are yellow, with the femora somewhat darker, and approaching the carapace in color. Length of female 2.5 mm; of male 2 mm.

This species has been collected from under dry leaves in sandy woods, as well as in deep ravines. It may be obtained by sifting forest floor litter, and has also been found in the nests of ants.

New England and adjacent Canada west to Utah.

16b   Metatarsus IV without a trichobothrium. Male palp with femur armed with teeth, and cymbium angulate. Spiracle (a) conspicuous and removed from base of spinnerets about one-third the distance to epigastric furrow (fig. 321) . . . .
. . . . ***Tennesseellum* (1 species)**

Figure   322.   *Tennesseellum   formicum* (Emerton)

The cephalothorax and legs are orange yellow. The abdomen is white with a gray band encircling it near the front and another at the rear. In some specimens the gray is quite indistinct on the dorsum though usually visible on the sides and venter. In others the abdomen is almost all gray. In both sexes a constriction or depression, more pronounced in males, is present at about the middle of the dorsum. Length 1.8 to 2.5 mm.

Among dead leaves on the forest floor.

New England south to Georgia and west to California.

FAMILY MICRYPHANTIDAE
Dwarf Spiders

The Micryphantidae, or dwarf spiders, (also known as the Erigonidae), is a very large family of spiders most of which are of small size, rarely over two millimeters. Because of this and the fact that most live under dead leaves and debris on the ground they are rarely seen by the casual collector. Only a few live in grass and on higher vegetation.

Many of the spiders in this family are provided with abdominal scuta, of variable extent (figs. 323 and 324), and in a number the head of the male is modified in one way or another. Some have elongate horns, some broad rounded protuberances, (fig. 325) and others tufts of bristles. In one group there are grooves or pits in the eye region (fig. 324) so that at first glance it appears as though there are more than eight eyes.

Many spider students consider this family as only a subfamily within the Linyphiidae. As indicated under that family (p. 115) it is often difficult to place properly in one or the other some of the members of these two groups. Almost any character selected will be found to have exceptions.

Classification is based almost entirely on the structure of the complicated male genitalia, and it is a difficult task even for the araneologist. It is beyond the scope of this book.

Figure 325. *Hypselistes florens*, face and chelicerae of male.

FAMILY THERIDIOSOMATIDAE
Ray Spiders

The members of this small family had long been placed in the Argiopidae (in the broad sense), but separable from the Araneidae, from which they differ as given in the key to families. There is one genus including three species.

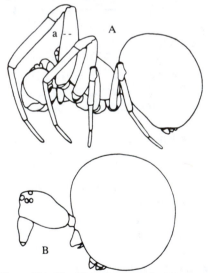

Figure 326. *Theridiosoma*. A, male; B, Female.

Figure 326. *Theridiosoma radiosa* (McCook). Ray Spider.

The color varies from a dirty yellow to gray and the abdomen is marked with silvery spots. Length of female 2.7 mm; of male 1.6 mm.

The web is different from the usual orb. The radii, instead of all converging on one center, are united in groups of three or four,

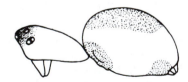

Figure 323. *Ceraticelus,* male from side.

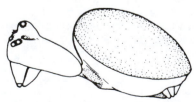

Figure 324. *Pelecopsis,* male from side.

each group connected with the center by a single thread (fig. 328). The web is drawn into the shape of a cone by a thread extending from the center to a neighboring twig and held tightly by the spider, who releases it suddenly to aid in entangling any insect that might chance to touch the snare.

Found in dark and damp situations on the banks of creeks, and in wet moss on the faces of cliffs. The cocoons (fig. 327) are almost spherical, brownish yellow, papery, and suspended by a thread.

New England and adjacent Canada south to Georgia and west to the Mississippi River.

Figure 327. Egg sacs of *Theridiosoma radiosa*.

Figure 328. *Theridiosoma* snare.

FAMILY ARANEIDAE

Typical Orb Weavers

This is a very large family, with 26 genera and probably several hundred species in our region. Many workers prefer to use the name Argiopidae, which, however, usually is considered to include the spiders which in this book are placed in the Theridiosomatidae and Tetragnathidae.

Almost all of our species spin snares in the form of an orb, and while some build a retreat at a distance from the snare, others remain at the center of the orb, hanging head down in most cases, quietly awaiting their prey.

In constructing the snare the spider first ties together the objects between which the web is to be spun. Often, this involves making use of air currents to carry the thread. The spider tests the line emitted into the wind to determine when it has stuck. The bridgeline is then pulled taut and fastened, and serves as a foundation thread. This remains in use for a long time although in some species the rest of the snare may be renewed every few days.

A frame is then spun with other foundation lines, which, with the bridge, determine the plane of the web. It may be vertical, horizontal, or any angle between. The first radius is passed through the center of the entire space (as a diameter). The spider crawls along this to where the center of the web will be and fastens at this point the line which will be the next radius. Running along the radius already present, the new thread is paid out from the spinnerets and held clear with a hind leg. When the foundation frame thread is reached she moves a certain distance along it, pulls the new thread taut and fastens it. She then returns to the center on this newest radius and in so doing strengthens it with a second thread, which is not held clear, but allowed to adhere, thus doubling the line.

The remaining radii are laid down in the same manner and all are fastened in the center so that a kind of mesh, the hub, is formed.

In order that the radii remain at the required distance from one another and that they sustain equally the stretching of the web they are connected by means of a few turns of a spiral thread spun just outside the hub. This is the attachment zone. Each time after spinning a radius the spider ascertains the tension of the web at the center in order to find in what direction the next radius should be spun. Therefore, they are not put in consecutively, but alternately on opposite sides of the web space. The number of radii may differ from species to species, but for a given species is fairly constant within certain limits.

Once the hub and attachment zone are completed the spider spins a spiral thread starting at the end of the attachment zone spiral and continuing to the periphery, with turns as far apart as the spider can stretch. The thread is not viscid and its function is to hold the radii in place during subsequent operations. The real snaring spiral of viscid thread is now begun. This is always spun from periphery to center, though usually not all the way to the attachment zone, leaving an area of greater or less width, called the free zone. The original holding spiral is cut away turn by turn as the viscid spiral approaches it.

Once the viscid spiral is finished the spider has only to make a change in the hub, which is the most variable part of the web. Some spiders bite out the threads from the center so that the hub becomes an open one; then the attachment zone is strengthened further. Others cover the hub with a dense sheet of silk. Some construct a band of silk below, or above, or both above and below the center. Thus, the stabilimentum, may be straight, or zigzag.

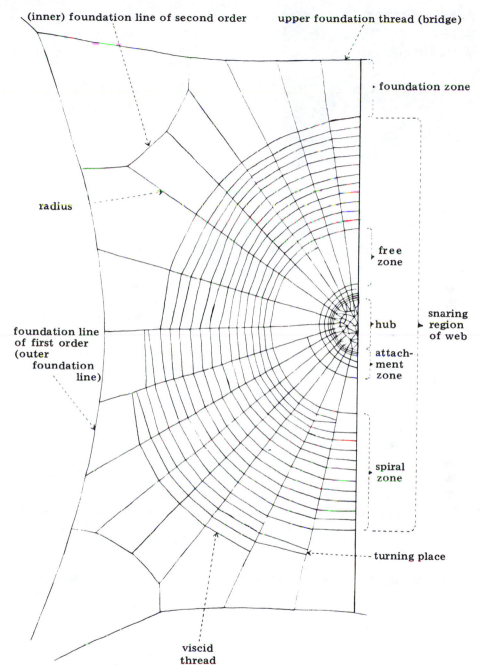

(inner) foundation line of second order

upper foundation thread (bridge)

foundation zone

radius

free
zone

hub

snaring
region
of web

foundation line
of first order
(outer
foundation
line)

attach-
ment
zone

spiral
zone

turning place

viscid
thread

Figure 329. Diagram of the web of an orb-weaving spider.

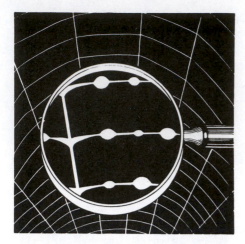

Figure 330. Portion of an orb web showing the viscid spiral threads magnified.

In many species there is considerable sexual dimorphism, with the males often being much smaller, provided with special clasping spines or spurs on the legs, and differing even in the shape of the cephalothorax as well as abdomen. They are not as frequently seen as the females.

1a Spinnerets elevated on a large projection and occupying a circular space limited by a thick flange in the form of a tube or ring (fig. 331). Abdomen hard, flattened on the dorsum and provided with conical humps and spines. Spiny-bellied Orb Weavers, Subfamily Gasteracanthinae. . 2

Figure 331. *Gasteracantha,* showing spinnerets surrounded by a sclerotized wall.

1b Spinnerets not surrounded by a sclerotized wall 3

2a Cephalothorax at least as wide as long; thoracic furrow procurved; abdomen transverse, much wider than long (fig. 332) . . . . *Gasteracantha* (1 species)

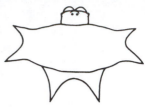

Figure 332. *Gasteracantha elipsoides,* female.

Figure 332. *Gasteracantha elipsoides* (Walckenaer)

This species is also known under the name *cancriformis* (Linnaeus). The carapace is dark brown. The abdomen is orange or yellow with the pointed spurs red, and with dark brown oval spots above. The venter is black with small yellow spots. Length of female 8 to 10 mm; of male 2 to 3 mm.

In wooded areas.

North Carolina south to Florida and west to California.

2b Cephalothorax longer than wide; thoracic furrow a rounded pit; abdomen more or less elongate . . . . *Micrathena* (5 species)

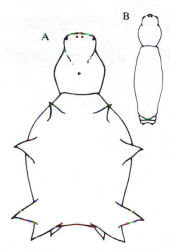

Figure 333. *Micrathena gracilis.* A, Female; B, Male.

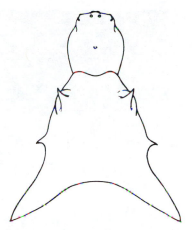

Figure 334. *Micrathena sagittata,* female.

Figure 333. *Micrathena gracilis* (Walckenaer)

The colors are white or yellow, and brown spotted, but there is much variation, some individuals being all white and others almost entirely black. There are five pairs of conical tubercles. The abdomen of the male is quite elongate and widest about at the middle of its length, with the spinnerets nearer the pedicel than the posterior end. Length of female 7.5 to 10 mm; of male 4.5 to 5 mm.

In dense wooded areas.

All states east of the Rockies.

Figure 334. *Micrathena sagittata* (Walckenaer)

There are three pairs of pointed tubercles on the abdomen, with the posterior pair widely spreading to make the abdomen triangular from above. These tubercles are black at the points and red at the base. The middle of the dorsum is bright yellow. In the male the abdomen is widest at the rear, without spines, and the spinnerets are nearer to this end than to the pedicel. Length of female 8 to 9 mm; of male 4 to 5 mm.

The web is built in open woods, often with a short stabilimentum above the hub (fig. 335).

Eastern states west to Texas and Nebraska.

Figure 335. *Micrathena sagittata,* snare.

3a   Epigastric plates over book lungs marked by transverse furrows (a) (fig. 336). Chelicerae usually with a well developed boss. 5

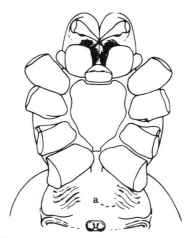

Figure 336. *Araneus* from below showing furrows on epigastric plates.

3b Epigastric plates without transverse furrows. Chelicerae with the boss rudimentary (see fig. 337B) . . . . **Subfamily Metinae 4**

4a Posterior femora with a double fringe of hairs on the prolateral surface of the basal half (fig. 337A) . . . . *Leucauge* (2 species)

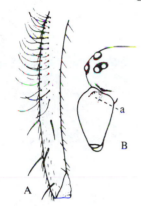

Figure 337A. *Leucauge,* femur IV.
Figure 337B. *Leucauge,* chelicera showing rudimentary boss, (a).

Figure 338. *Leucauge venusta.*

Figure 338. *Leucauge venusta* (Walckenaer). Orchard Spider.

The carapace is yellowish green with darker stripes on the sides. The abdomen is elliptical, silvery above, and with dark lines. The sides are yellow with red spots near the posterior end, and there is a red spot in the middle of the venter. Length of female 5.5 to 7.5 mm; of male 3.5 to 4 mm.

In low bushes and trees.

New England south to Florida and west to Texas and Nebraska.

4b Posterior femora not so fringed . . . . *Meta* (1 species)

Figure 339. *Meta menardii* abdomen.

Figure 339. *Meta menardii* (Latreille) Cave Orb Weaver.

The carapace is yellowish to brown with a deep pit at the dorsal groove. The abdomen is dark brown to purple with a pattern of light areas as shown. Length of female 8 to 10 mm; of male 8 to 9.5 mm.

In caves, deserted mines, and heavily shaded ravines.

New England and adjacent Canada south to Virginia and west to the Mississippi.

5a On each leg the tarsus and metatarsus together longer than the tibia and patella together........................ 6

5b Tarsus and metatarsus together not longer than tibia and patella together........ .... Subfamily Araneinae (in part) 9

6a Labium longer than broad. Abdomen two and one-half to three times as long as broad. Sustentaculum lacking from tarsus IV. .... Subfamily Nephilinae, Genus *Nephila* (1 species)

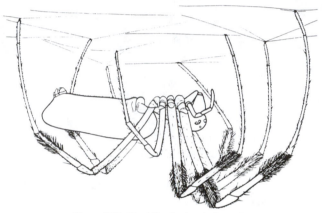

Figure 340. *Nephila clavipes* in its web.

Figure 340. *Nephila clavipes* (Linnaeus). Golden-silk Spider.

This species is readily recognized by the conspicuous tufts of hairs on femora and tibiae of legs I, II, and IV. The carapace is dark brown,

the abdomen olive green with pairs of yellow and white spots. There is also a light line across the anterior end of the abdomen. Length of female 22 mm; of male 5 to 8 mm.

The webs are large and built in shaded woods. They differ in a number of respects from others in the family. The radii are pulled out of their direct course to give a notched appearance (fig. 341) and the viscid spiral is of yellowish color rather than white.

Southeastern States.

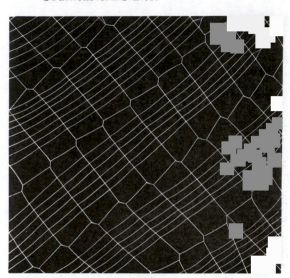

Figure 341. Portion of *Nephila* web, showing detail.

6b Labium broader than long. Abdomen not longer than two and one-half times its width. Sustentaculum present on tarsus IV (fig. 342)..................... 7

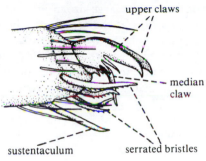

Figure 342. *Araneus*, tip of tarsus IV.

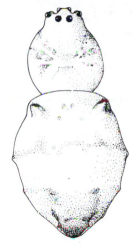

Figure 343. *Gea heptagon.*

7a **Posterior row of eyes strongly procurved. Anterior lateral eyes smaller than posterior laterals. . . . .**
. . . . Subfamily Argiopinae 8

7b **Posterior row of eyes straight or recurved. Lateral eyes subequal . . . .**
. . . . Subfamily Araneinae (in part) 9

8a **The eyes of the anterior row are either equidistant, or else the medians are closer to the laterals than to each other. Length of adult female less than 6 mm, and the abdomen is about 1.5 times the length of the cephalothorax. Tibia I of male curved and with many bristles. . . . .**
. Gea (1 species)

Figure 343. *Gea heptagon* (Hentz)

The carapace is yellowish, with darker brown markings between the radial furrows. The legs are yellowish with brown annuli. The abdominal dorsum is pale yellow, but with darker spots and a posterior dark triangle as shown in the illustration. Length of female 4.5 to 5.8 mm; of male 2.6 to 4.3 mm.

As in the case of *Argiope* the spider sits in the center of its snare, which occasionally has a sector missing from the *lower* half. (Compare with *Zygiella* where the sector is missing from the *upper* half.) The snare is built close to the ground in low vegetation and there is no stabilimentum. When frightened the spider drops from its web and displays a sudden darkening of the light areas, making it more difficult to see against the ground.

New Jersey south to Florida, west to Michigan, Kansas, and Texas and then to California.

8b **Anterior median eyes closer to each other than to the laterals. Length of adult female more than 9 mm, and the abdomen is two or more times the length of the carapace. Tibia I of male straight and unmodified . . . .**
. . . . Argiope (5 species)

GENUS ARGIOPE
Garden Spiders

These are large orb-weavers which build no retreat but are found sitting in the center of the snare, which is usually provided with a stabilimentum. While the stabilimentum is a characteristic feature there are times when the spider omits this feature from the web. The factors involved are unknown, and the exact function of the stabilimentum is uncertain.

Because of its large size, bright colors and its habit of sitting at the center of its web built in open, sunny places, this spider is one of the most commonly known. It builds webs in gardens around houses, and in tall grass. A zig-zag stabilimentum extends above and below the hub. The egg sacs are conspicuous spheroids narrowed at one end, about 20 to 25 mm long, and with a brownish, tough, papery cover (fig. 345).

Throughout the United States, but apparently not common in the Rockies and Great Basin areas.

Figure 345. Egg sac of *Argiope aurantia*.

Figure 344. *Argiope aurantia*, female.

Figure 344. *Argiope aurantia* Lucas. Black and Yellow Garden Spider; Writing Spider.

The carapace is covered with silvery hairs. The abdomen is slightly pointed behind and notched in front to form a hump on each side, and is marked in black and yellow (or orange). The front legs are sometimes black and sometimes with a short band of orange on the femur. The others have the femora reddish or yellow and the rest black. Length of female 19 to 28 mm; of male 5 to 8 mm.

Figure 346. *Argiope trifasciata*, female.

Figure 346. *Argiope trifasciata* (Forskal). Banded Garden Spider.

The ground color is pale yellow, with silvery hairs on the carapace and with numerous thin silver and yellow transverse lines alternating with black on the abdomen. The legs are spotted. The abdomen is usually more pointed behind than in *aurantia* and lacks the notch and humps in front characteristic of that species. Length of female 15 to 25 mm; of male 4 to 5.5 mm.

The webs are built in tall grass in sunny areas. The stabilimentum is similar to that of *aurantia* but not quite as prominent. The egg sacs are shaped like kettle drums (fig. 347) and are provided with a brown, tough, papery cover.

Throughout the United States.

Figure 347. Egg sac of *Argiope trifasciata*.

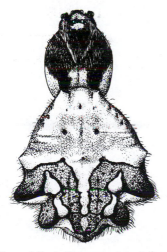

Figure 348. *Argiope argentata*, female.

Figure 348. *Argiope argentata* (Fabricius)

This species differs from the preceding two in having the abdomen relatively wider, up to nine-tenths its length at the widest point. Also the posterior end is lobed, a median and three pairs of lateral lobes giving a crenated appearance to the posterior end. The anterior half of the abdomen is yellowish, and the posterior half has a dark triangle as shown in the illustration. The carapace and most of the dorsum are covered with silvery hairs. Length of female 12 to 16 mm; of male 3.7 to 4.7 mm.

The web is fitted with a double stabilimentum so arranged as to make a St. Andrew's cross with the spider sitting where the two bands cross. In southern California the web is often found among cactus plants. The egg sac (fig. 349) is somewhat flattened from side to side and drawn out to several points along its edges. It is usually fashioned of silk with a brownish to greenish tinge.

Southern Florida west to southern California.

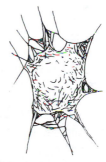

Figure 349. *Argiope argentata*, egg sac.

9a Tibia III bearing on its prolateral surface a double series of long thin feathery hairs (fig. 350). Cephalothorax with thoracic part higher than cephalic. (Palpal patella of male with only one dorsoapical spine) . . . . *Mangora* (4 species)

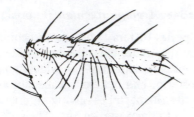

Figure 350. *Mangora*, tibia III.

Figure 351. *Mangora placida.*

Figure 351. *Mangora placida* (Hentz)

The carapace is yellow with a brown stripe in the middle and along each side. The abdomen is white either side of a dark brown median stripe, which is wider behind, and the sides are black. Length of female 2.3 to 4.5 mm; of male 2 to 2.8 mm.

In bushes and trees in wooded areas, and also in tall grass. The snare is very finely meshed.

Eastern states and adjacent Canada west o North Dakota and Texas.

Figure 352. *Mangora gibberosa.*

Figure 352. *Mangora gibberosa* (Hentz)

The carapace and legs are yellow to pale green with a thin black line in the middle. There is a similar line under femora I and II. The abdomen is yellow to light greenish gray with a pair of longitudinal black lines on the posterior half and several black spots in front. Length of female 3.4 to 6 mm; of male 2.6 to 3.2 mm.

It builds a very finely meshed snare (fig. 354) in low bushes and tall grass.

Eastern states and adjacent Canada west to North Dakota and Texas.

Figure 353. *Mangora maculata.*

Figure 353. *Mangora maculata* (Keyserling)

Very similar to *gibberosa* but lacks lines under the femora, on the carapace and on the abdomen. There are three pairs of spots on the posterior portion of the dorsum. Length of female 3.6 to 5.5 mm; of male 2.7 to 4 mm.

In low bushes in wooded areas.

Eastern states west to Nebraska and Texas.

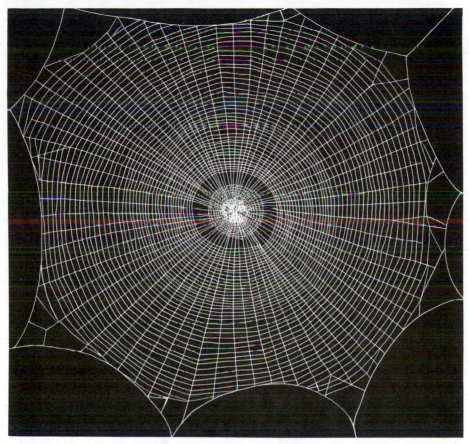

Figure 354. *Mangora gibberosa* snare.

9b **Tibia III without such hairs. Thoracic part not higher than head region 10**

10a **Approximately the basal two-thirds to five-sixths of the abdomen covered by a shiny scutum, which, however, does not obscure the pattern of markings. The anterior end of the abdomen overhangs the carapace with a blunt point, on either side of which are three to five spines set on dark elevations (fig. 355) . . . . *Cercidia* (1 species)**

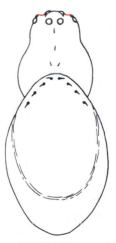

Figure 355. *Cercidia prominens*.

Figure 355. *Cercidia prominens* (Westring)

The body is brown to red with darker spots along the sides of the carapace and a suggestion of transverse black lines on the posterior half of the abdomen. The male has two apical dorsal spines on the palpal patella and the abdominal scutum is proportionately larger than in the female. Length of female 3.8 to 5 mm; of male 3 to 4.1 mm.

New England west to Wisconsin.

10b   Abdomen without scutum 11

11a   Abdomen hard and with a thick median cone-like protuberance at the base as well as lateral and posterior cones (fig. 356) . . . . . . *Acanthepeira* **(3 species)**

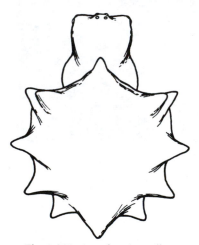

Figure 356. *Acanthepeira stellata.*

Figure 356. *Acanthepeira stellata* (Marx). Starbellied Spider.

The carapace is brown, the legs yellow annulated with brown. The abdomen is brown, lighter on the posterior two-thirds and with dark spots. On the median tubercle that overhangs the carapace there is a white spot. The male has two apical dorsal spines on the palpal patella. Length of female 7 to 15 mm; of male 5.1 to 8.1 mm.

In tall grass and low bushes.

New England and adjacent Canada south to Florida and west to Kansas and Arizona.

11b   Abdomen elongate, or oval, or triangular, but without the cones as illustrated; with at most fewer and smaller tubercles . . . 12

12a   Abdomen triangular ovate above, flattened on top, with one or more tubercles at the posterior end, or if the posterior tubercle is not conspicuous, then the posterior end is distinctly truncate as in figure 357 13

12b   Abdomen not flat above, and not triangular-ovate. 15

13a   Scape of epigynum thin and long, and reaching posteriorly almost to the spinnerets (figs. 359B and 360) 14

13b   Scape short, and with its free and extending forward (Males with palpal patella bearing only one dorsoapical spine.). . . . .
. . . . *Eustala* **(4 species)**

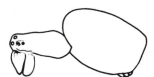

Figure 357. *Eustala anastera.*

Figure 358. *Eustala anastera.*

Figure 358. *Eustala anastera* (Walckenaer)

The carapace is gray, darker at the sides. The abdomen shows a considerable variety of patterns. Most commonly it is gray with a central triangle having scalloped edges. Length of female 5.4 to 8 mm; of male 4 to 6 mm.

In low trees and among shrubs and bushes.

New England to Florida and west to the Rockies.

Eustala rosae Chamberlin & Ivie
Similar to the preceding species, but with three tubercles at the posterior end, instead of just the one. Length of female 9.5 mm; of male 4 to 6.2 mm.

Texas west to California.

14a   Integument on dorsum hard and glossy. Hind end with three pairs of short tubercles, each pair with an upper anterior and a lower posterior member...
. . . . *Verrucosa* (3 species)

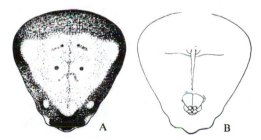

Figure 359. *Verrucosa arenata*. A, Abdomen from above; B, from below.

Figure 359. *Verrucosa arenata* (Walckenaer)

The elevated head is much darker brown than the thoracic portion. The abdomen is almost completely covered with a white, yellow, or pink triangle. Length of female 8 to 9 mm; of male 5.5 to 6 mm.

In deciduous forests. Contrary to the usual case this spider sits at the center of the snare with its head end *up!*

New York south to Florida and west to Kansas and Texas.

14b   Integument not hard and glossy. Hind end with only a single median tubercle (as well as a shoulder hump each side). . . . . .
. . . . *Eriophora* (2 species)

Figure 360. *Eriophora edax*, side view of abdomen showing long scape of epigynum.

Figure 361. *Eriophora edax*, abdomen from above.

Figures 360 and 361. *Eriophora edax* (Blackwall)

This is an exceedingly variably marked species. In some specimens the abdomen above is dark with a mottling of light. In others it is light with a small dark diamond-shaped black spot up front. In others it is orange tan with a thin white stripe along the mid-dorsum. Some are evenly orange pink all over. The venter shows a large black triangle, its apex pointing to the rear. The caudal tubercle is conspicuous. Length of female 12 to 16 mm; of male 11 to 13 mm.

Southern Texas west to southern California.

15a   Abdomen elongate oval to rhomboidal with a triangular folium having un-

dulating margins bordered by a white line
(fig. 362) . . . .
. . . . *Acacesia* (1 species)

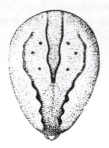

Figure 362. *Acacesia hamata.*

Figure 362. *Acacesia hamata* (Hentz)

The carapace and legs are greenish gray to
brown. The abdomen has a ground color of
gray to green with black and white lines as
figured. Length of female 4.7 to 9.1 mm; of
male 3.6 to 5 mm.

In bushes and shaded woods.

New England south to Florida and west
to Illinois and Texas.

**15b Abdomen otherwise, with a different
pattern. 16**

**16a In both sexes, on each leg the tarsus plus
metatarsus is longer than the patella plus
tibia. 17**

**16b Tarsus plus metatarsus not longer than
tibia plus patella (except on legs I and II
of some males). 19**

**17a Abdomen ovate, not produced over
carapace in front (fig. 363). Venter with a
median longitudinal broad white line. . . .
. . . . *Metepeira* (16 species)**

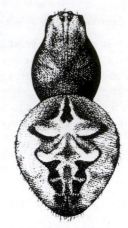

Figure 363. *Metepeira labyrinthea.*

Figure 363. *Metepeira labyrinthea* (Hentz)

The carapace is brown, much lighter in the eye
region. The abdomen is brown with a distinct
folium of white and black areas as figured.
Length of female 5.5 to 6.2 mm; of male 4 to
4.5 mm.

The snare is a composite, consisting of a
more or less vertical orb behind which is placed
an irregular labyrinth resembling the snare of a
theridiid (fig. 364). Often a small silken tent is
constructed in the mesh, and this serves as a
retreat.

Eastern States to Wisconsin.

Metepeira arizonica Chamberlin & Ivie
Similar in general appearance to *labyrinthea.*
Length of female 6.2 to 7.2 mm; of male
4.2 mm.

In the arid regions of Arizona and
California.

Metepeira crassipes Chamberlin & Ivie
Similar to *labyrinthea* in general appearance.
Length of female 5.9 to 7.2 mm; of male 2.8 to
3.8 mm.

California.

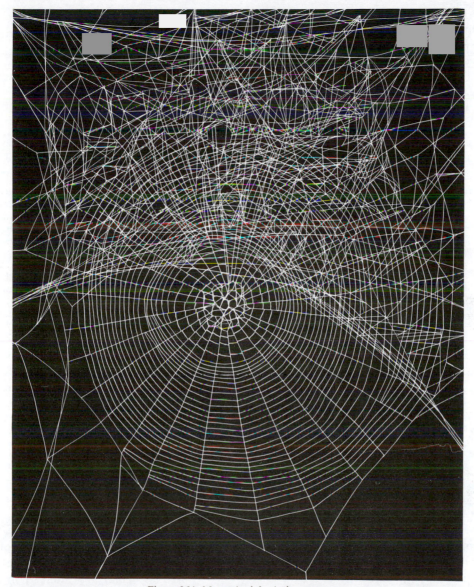

Figure 364. *Metepeira labyrinthea* snare.

Metepeira foxi Gertsch & Ivie
Similar to the preceding species. Length of female 6 mm; of male 3.5 mm.

Rocky Mountain States west to California.

17b **Abdomen elongate, at least twice as long as wide, and produced in front where it overhangs the carapace. Venter with two white lines or none**.................**18**

18a Sternum at least one and two-thirds as long as wide. Endites much longer than wide. Median ocular area about twice as wide in front as behind. Abdomen with a single median blunt point in front (fig. 365). Males with palpal patella bearing two heavy spines at apex . . . .
. . . . *Larinia* (3 species)

Figure 365. *Larinia directa.*

Figure 366. *Larinia borealis,* female.

Figure 366. *Larinia borealis* Banks

The carapace is yellow with a dark line extending back from each posterior median eye. The abdomen is yellow with a pair of pale gray bands, somewhat darker posteriorly, running the length of the dorsum. On the venter three black lines alternate with two yellow. The abdomen is twice as long as wide. Tibia plus patella I of the female is 1.2 to 1.6 times as long as the carapace; of the male is 1.4 to 1.7 times as long. In both sexes the length of tibia plus patella I is about the same as metatarsus plus tarsus I. Length of female 4.5 to 8 mm; of male 3.9 to 5.4 mm.

New England and adjacent Canada south to Virginia and west to California and eastern Washington.

Figure 365. *Larinia directa* (Hentz)

The carapace is yellow with a faint brown line extending back from the posterior median eyes. The abdomen is orange yellow, quite variable in markings, but usually with indications of four pale red longitudinal lines. Others have two rows of six or more black spots and a single black spot on the blunt cone overhanging the carapace in front. The specimens with spots on the abdomen also have the legs spotted. The abdomen is more than twice as long as wide. Tibia plus patella I of the female is 1.6 to 2.2 times as long as the carapace; of the male is 2 to 2.4 times as long. In both sexes the tibia plus patella I is less than the length of metatarsus plus tarsus I. Length of female 4.8 to 11.7 mm; of male 4.5 to 6.5 mm.

In grass in sunny areas.

New Jersey south to Florida and west through the Southern States to California.

18b Sternum at most only slightly longer than wide. Endites almost as wide as long. Median ocular area almost, or quite, as wide behind as in front. Abdomen with a hump on either side of the blunt point in front (fig. 367). Males with palpal patella bearing one heavy and one lighter spine at apex . . . .
. . . . *Mecynogea* (1 species)

Figure 367. *Mecynogea lemniscata.*

Figure 367. *Mecynogea lemniscata* (Walckenaer). Basilica Spider.

This species is also known as *Allepeira conferta.* The carapace is yellow with a narrow black line in the middle and wider ones at the margins. The abdomen is olive green with a black and yellow folium outlined in white. Length of female 6 to 9 mm; of male 5 to 6.5 mm.

The snare is an orb pulled into a dome, and as in *Metepeira,* also provided with a labyrinth adjacent. The egg sacs are attached to each other in strings and suspended in the web (fig. 368).

District of Columbia south to Florida and west to Colorado.

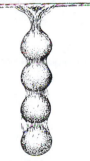

Figure 368. Egg sacs of *Mecynogea lemniscata.*

19a Abdomen with a caudal tubercle (fig. 369) (reduced in males). Eyes on prominent tubercles (Males with palpal patella bearing only one dorsoapical spine) . . . . *Cyclosa* (3 species)

Figure 369. *Cyclosa conica,* female.

Figure 369. *Cyclosa conica* (Pallas)

The color varies from a mixture of gray and white to an almost all black. Light individuals are darker in the middle of the dorsum. The caudal hump varies in size in different individuals becoming more prominent with age, and reduced in males. Length of female 5.3 to 7.5 mm; of males 3.6 to 4 mm.

The snare is built in open woods, and is usually provided with a short stabilimentum above and below the hub (fig. 371).

Throughout the United States.

19b Abdomen without such a caudal projection 20

20a Thoracic groove longitudinal (fig. 370) 21

Figure 370. *Neoscona carapace.*

Figure 371. *Cyclosa conica* snare.

20b **Thoracic groove transverse, straight or recurved, or a circular pit (fig. 372) . . . 23**

Figure 372. *Araneus* carapace.

21a **Abdomen as high behind middle of its length as at base and elliptical in outline. (Posterior median eyes smaller than the anterior medians). 22**

21b **Abdomen higher near anterior end and oval or triangular oval in outline. (Lateral eyes not on distinct tubercles) . . . . *Neoscona* (7 species)**

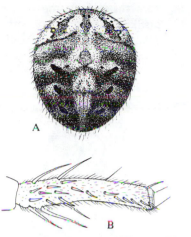

Figure 373. *Neoscona arabesca*. A, Abdomen of female; B, Right tibia II of male from below.

Figure 373. *Neoscona arabesca* (Walckenaer)

The colors are yellow and brown with paired black spots on the posterior half of the dorsum and lighter areas as figured. Length of female 5 to 12.3 mm; of male 4.2 to 9.2 mm. Tibia II (B) of the male is curved.

The webs are built in tall grass and low bushes.

Throughout the United States and southern Canada.

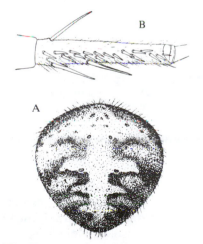

Figure 374. *Neoscona domiciliorum*. A, abdomen of female; B, right tibia II of male from below.

Figure 374. *Neoscona domiciliorum* (Hentz) and *N. sacra* (Walckenaer)

These two species are very similar and are easily confused, and they had long been considered as one, *benjamina* (Walckenaer). The markings are slightly more pronounced in *domiciliorum*, less so in *sacra*, which latter is by some considered to be *hentzi* (Keyserling). The abdomen is more triangular-ovate than in *arabesca*. Tibia II of the male is straight. Length of female for *domiciliorum* 7.2 to 16.2 mm; of male 6 to 9 mm; for *sacra* female 8.5 to 19.7 mm; of male 4.5 to 15 mm.

The web is built in open woods, but also around fences and buildings. *N. domiciliorum* is found from New England south to Florida and west to Indiana and Texas. *N. sacra* is more common in the north but its range extends into Florida and west to Arizona, so there is some overlap.

Figure 375. *Neoscona oaxacensis*.

Figure 375. *Neoscona oaxacensis* (Keyserling)

The colors are yellowish with brown markings. The carapace has a dark median stripe and a dark stripe along each lateral margin. The dorsum shows a median yellowish herringbone pattern and a mottling of light spots to the sides of this. Length of female 11 to 17 mm; of male 5 to 12 mm.

Texas north to Kansas and west to California.

22a Median ocular area much wider in front than behind. Venter and sternum unmarked. Patella plus tibia I about as long as carapace . . . .
. . . . *Metazygia* (1 species)

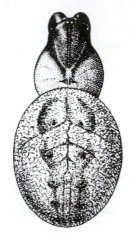

Figure 376. *Metazygia wittfeldae.*

Figure 376. *Metazygia wittfeldae* (McCook)

The carapace is orange brown in front and yellow behind. The abdomen is yellow with a brown folium as figured. The female lacks dorsal spines on tibia I and II. The male, however, has one dorsal spine on each tibia and has a single large spine on the palpal patella. Length of female 8 to 10 mm; of male 6 to 7 mm.

In tall grass and on bridges.
Florida west to Texas.

22b Median ocular area as wide behind as in front. Venter and sternum with dark bands. Patella plus tibia I longer than carapace . . . .
. . . . *Zygiella* (5 species)

GENUS ZYGIELLA

Our species resemble one another closely in general appearance. The webs usually have the viscid spiral threads missing from a definite sector between two radii in the upper part of the web, and in this open area a strong signal line extends from the hub to the spider's retreat, which is a silken tube open at both ends. The spider is seen on the web itself only at night. Young spiders are more apt to build complete orbs, and sometimes even adults will build a web without a sector missing.

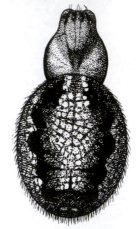

Figure 377. *Zygiella nearctica.*

Figure 377. *Zygiella nearctica* Gertsch

The carapace is yellowish with a thin black marginal line. Abdomen yellow to brown with a black bordered folium as figured. Length of female 6 to 7 mm; of male 5 to 6 mm.

New England south to North Carolina and west to the Rockies.

Zygiella x-notata (Clerck)
Similar to *nearctica* with a gray folium on the dorsum. Length of female 7.4 to 8.7 mm; of male 6 to 6.5 mm.

New England south to Virginia and west to the Pacific Coast States and British Columbia.

Zygiella atrica (C.L. Koch)
Similar in appearance to *nearctica* and *x-notata*. The male is easily distinguished by the fact that the pedipalp is as long as the entire body. It is much shorter in the other two species. Length of female 8 mm; of male 5.5 mm.

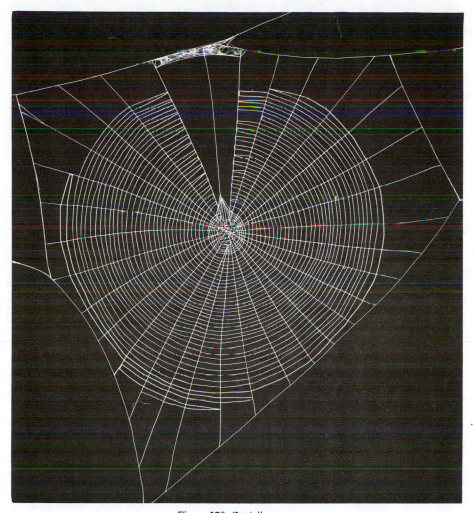

Figure 378. *Zygiella* snare.

This species tends to build its web in moister situations than does *x-notata*.

New England, New York and eastern Canada; also British Columbia.

23a   Abdomen usually as high behind middle of its length as at base and elliptical in outline. Integument shiny. Legs short, with patella plus tibia I hardly as long as carapace........................ 24

23b   Abdomen usually higher near anterior end and usually oval, or triangular oval in outline. Integument not especially shiny. Legs not so short, with tibia plus patella I about one and one-half or more times as long as carapace. (Lateral eyes often on distinct tubercles)................. 26

24a Median ocular area not wider in front than behind. Height of clypeus 1.5 to 3 times the diameter of an anterior median eye. Posterior median eyes larger than the other eyes.......................
.... *Hypsosinga* (5 species)

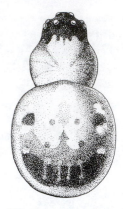

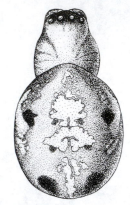

Figure 379. *Hypsosinga pygmaea*, female.

Figure 380. *Hypsosinga rubens*, female.

Figure 379. *Hypsosinga pygmaea* (Sundevall)

The carapace is light orange with black in the eye region. The legs are orange without any markings. The abdomen is exceedingly variable, which explains the other name under which it is known, *variabilis* (Emerton). In many the abdomen is all black, in others there are lighter markings. The figure shows one variety. Length of female 2.9 to 4 mm; of male 2.2 to 3.1 mm.

New England south to Florida and west to Washington. Also throughout Canada to Alaska.

Figure 380. *Hypsosinga rubens* (Hentz)

This species is also known under the name *truncata* (Banks). The carapace is orange with black around the eyes, the black areas being more extensive in females than in males. The legs are orange, and somewhat dusky on the more distal segments. The abdomen may be orange without markings, or with gray to black near the posterior end of the dorsum. Length of female 2.4 to 5.1 mm; of male 2.7 to 3.1 mm.

New England south to Florida and west to South Dakota and Texas. In Canada from Nova Scotia west to Alberta.

24b Median ocular area wider in front than behind. Height of clypeus equal to, or less than, the diameter of one of the anterior median eyes, which are larger than the other eyes 25

25a Epigynum with a scape. Height of clypeus equal to the diameter of an anterior median eye ....
.... *Singa* (2 species)

Figure 381. *Singa keyserlingi,* female.

Figure 381. *Singa keyserlingi* McCook

The carapace is orange with black in the eye region, and slightly posterior to them, more extensive in females than males. The abdominal dorsum shows two wide black bands alternating with three white bands in the female. In the male it is all black. The legs are orange. Length of female 5.1 to 6 mm; of male 2.3 to 4 mm.

New York southwest to Alabama and northwest to Montana. In Canada from Ontario to Alberta.

25b Epigynum lacking a scape. Height of clypeus less than the diameter of an anterior median eye.................
..................*Alpaida* (1 species)

Figure 382. *Alpaida calix.*

Figure 382. *Alpaida calix* (Walckenaer)

This species is also known under the name *maura* (Hentz). The cephalothorax and legs are orange. The abdomen is brown to orange, lighter on the sides and with white areas along the middle. At the posterior end is a pair of large oval black spots. Length of female 3.9 to 6 mm; of male 3.8 to 4 mm.

From Maryland south to Florida and west to Illinois.

26a Sternum not, or scarcely, as long as wide. Cephalic region not very sharply set off from the thoracic. Tibia II of female lacking spines beneath ....
.............*Neosconella* (5 species)

Figure 383. *Neosconella pegnia.*

Figure 383. *Neosconella pegnia* (Walckenaer)

The color is yellow to brown, with two pairs of large white spots on the anterior half of the abdomen and dark brown to black areas on the posterior half. Length of female 3.5 to 8.2 mm; of male 2.5 to 5 mm.

The snare includes an orb, with sometimes a sector free of viscid threads (as in *Zygiella*), and an irregular net similar to that used by *Metepeira*.

Throughout the United States.

Neosconella thaddeus (Hentz). Lattice spider.
The pattern is variable, with some specimens very closely resembling *pegnia,* and others having the folium completely lacking so that the dorsum is all white. Length of female 5 to 8 mm; of male 3.7 to 5.7 mm.

New England south to Georgia and west to Utah and Arizona.

26b Sternum longer than wide. Cephalic region sharply set off from thoracic by a deep groove. Tibia II of female with ventral spines 27

27a Median ocular area as wide behind as in front and the posterior medians slightly larger than the anterior medians. The tibia of the male palp has three spines . . .
. . . . *Araniella* (1 species)

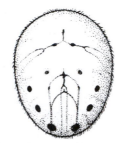

Figure 384. *Araniella displicata.*

Figure 384. *Araniella displicata* (Hentz)

The carapace and legs are yellow to brown without markings. The abdomen is white, or yellow or pink, with three pairs of black spots on the posterior half. Length of female 4 to 8 mm; of male 4 to 6 mm.

In tall grass and bushes.

New England south to North Carolina and west to Minnesota; also the Rockies west to the Pacific Coast States, and all of Canada to Alaska.

27b Median ocular area wider in front than behind; the posterior medians not larger than the anterior median eyes. The tibia of the male palp has one or two spines . 28

28a Abdomen somewhat flattened dorsoventrally and widest about the middle of its length. Venter black in the middle and with a white comma-shaped mark each side. Carapace setose. (Shoulder humps lacking; dorsum with a black or brown folium and a dark cardiac mark). . .
. . . . *Nuctenea* (3 species)

GENUS *NUCTENEA*

These spiders are often found around houses and other man-made structures. Mature individuals of both sexes can be found at all seasons of the year, and the males are able to mate three or four times with each palp. Webs are spun in the evening and usually contain fewer than 20 radii, and have the spiral turns widely separated so that the entire web shows an ''open'' appearance.

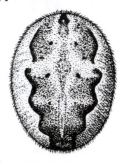

Figure 385. *Nuctenea cornuta.*

Figure 385. *Nuctenea cornuta* (Clerck) Furrow Spider.

This species is also known under the name *foliata* (Fourcroy). It and the two following are very similar in appearance. This one is the most common and the lightest of the three. The carapace is brown to gray. The abdomen has a folium which is grayish brown, somewhat darker than the sides, and with lighter areas within. The edges of the folium are usually entire, but sometimes broken. The front legs are shorter than in the other two species. Length of female 6.5 to 14 mm; of male 4.7 to 9 mm.

Throughout the United States, but more common east of the Mississippi River and in the more northern States. Also, all of Canada northwest to Alaska.

Figure 386. *Nuctenea sericata.*

Figure 386. *Nuctenea sericata* (Clerck). Bridge Spider; Gray Cross Spider.

This species is also known under the names *sclopetaria* (Clerck) and *undata* (Olivier). The general hue is darker and more gray than in *cornuta* and *patagiata*. The folium is dark brown to black, with yellowish areas enclosed, and the lateral edges are broken just anterior to the middle of its length. The front legs are longer than in the other two species. Length of female 8 to 14 mm; of male 5.5 to 8.5 mm.

The webs are often found under bridges and culverts, etc. A special retreat is not built, but the spider merely waits at the end of one of the foundation lines or radii.

New England and adjacent Canada south to Virginia and west to the Mississippi River. Also the Pacific Northwest north to Alaska.

Figure 387. *Nuctenea patagiata.*

Figure 387. *Nuctenea patagiata* (Clerck)

This species is also known under the name *dumetorum* (Villers) and *ocellatus* (Clerck). The general color is often reddish brown, with the folium darker than in *cornuta*. The front legs are intermediate in length and thickness between *cornuta* and *sericata*. Length of female 5.5 to 11 mm; of male 5 to 7 mm.

A special signal thread may be used, but as a rule the spider enters or leaves the web by any one of the ordinary radii.

New England south to North Carolina and west to the Pacific Coast States. Also all of Canada northwest to Alaska.

28b Abdomen not flattened, in most, highest at the front and widest anterior to the middle. Venter not as above and carapace not setose. (Shoulder humps present in some. Folium present or not, although most are brightly colored) . . . . *Araneus* **(49 species)**

GENUS ARANEUS

This genus contains our largest specimens of the subfamily Araneinae. Some (e.g., *diadematus, gemma, nordmanni,* and *saevus*) show well developed shoulder humps and are often referred to as "angulate" forms. Others, without these humps, are referred to as the "round-shouldered" forms. The males can be found only during brief periods of maturity. Each palp can be used only one time for mating. Many of the smaller species have been taken from the mud nests of sphecid wasps.

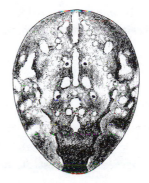

Figure 388. *Araneus diadematus.*

Figure 388. *Araneus diadematus* Clerck. Cross Spider; Garden Spider (of Europe).

The general color varies from a pale yellow brown to nearly black. The folium is not generally as distinct as in some of the other species, and includes within it a number of white or yellow spots, the largest of which are arranged longitudinally near the anterior end. There is usually a pair of white spots at right angles to the longitudinal ones, giving the group the form of a cross (hence the common name). This cruciform arrangement of spots is more distinct in the darker varieties. The carapace has a median as well as marginal dark bands. Length of female 6.5 to 20 mm; of male 5.5 to 13 mm.

The spider may be found sitting in the center of its web.

New England and adjacent Canada across the northern States to Washington, Oregon, and British Columbia.

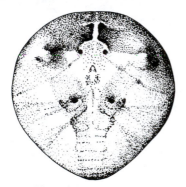

Figure 389. *Araneus gemma.*

Figure 389. *Araneus gemma* (McCook)

This is *one* of the largest, if not *the* largest of our "angulate" species. There is much variation in the intensity of pigmentation, with some of the darker specimens approaching the general appearance of *andrewsi*. The shoulder humps are very conspicuous, with a black spot just anterior to them. In many specimens these two spots are joined across the mid-dorsum with dark pigment. The rest of the dorsum is

yellow to tan, with an indication of a brownish, somewhat triangular, folium on the posterior half. But some specimens are quite light, with a very inconspicuous folium. Some have a narrow, whitish, parallel-sided, longitudinal band extending down the middle. The venter always shows a large dark spot, somewhat triangular in shape, pointing toward the spinnerets. On either side of this dark spot, near its posterior end is a small white spot. The scape of the epigynum is short and wide, its length less than the width of the epigynum. The legs are furnished with quite conspicuous annuli. Length of female 9 to 25 mm; of male 5.8 to 8.5 mm.

The spider builds its webs in bushes and trees, and around buildings, with a retreat quite close to the snare itself, and under a branch or leaf.

Idaho, Montana and the Pacific States north to British Columbia and Alaska.

Araneus andrewsi Archer

This is another very large "angulate" species, and in areas where both it and *gemma* occur, may be confused with the latter. However, it is generally darker, reddish brown to almost black. Consequently the folium and leg annuli are far less conspicuous. Furthermore, it does not show the ventral black and white areas described for *gemma*. The scape length is twice the width of the epigynum. Length of female 11 to 22 mm; of male 8 to 11 mm.

The web is built with a retreat quite a distance from the snare itself, deep in the bush.

California and Oregon.

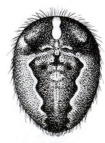

Figure 390. *Araneus nordmanni.*

Figure 390. *Araneus nordmanni* (Thorell)

The colors are brown to gray with white and yellow lines and spots. The shoulder humps are not as conspicuous as in the preceding three species. The pattern is quite variable, and one form is illustrated here. Length of female 7 to 19 mm; of male 5 to 10 mm.

New England and adjacent Canada south to North Carolina and west to Michigan; also the Rockies and west to the Pacific Coast States, British Columbia and Alaska.

Figure 391. *Araneus saevus.*

Figure 391. *Araneus saevus* (L. Koch)

This species is also known under the name *solitarius* (Emerton). This is similar to *nordmanni* but is darker. Length of female 11 to 21 mm; of male 9 to 14 mm.

New England across the northern States to Washington and Oregon.

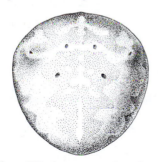

Figure 392. *Araneus cavaticus,* female.

Figure 392. *Araneus cavaticus* (Keyserling)

This is another angulate species. The general hue varies from a pale bluish gray to a grayish brown. The legs of the male are very long and thin, and are densely covered with long thin spines. Length of female 13 to 22 mm; of male 10 to 19 mm.

The webs are often found on bridges, barns, and other man-made structures, as well as from overhanging cliffs.

New England and adjacent Canada southwest through West Virginia to Alabama and Texas.

Araneus gemmoides Chamberlin and Ivie
The carapace is light brown. The abdomen is light brown to gray with the females hardly showing any pattern, although the folium is somewhat more distinct in males. Like *cavaticus* it is often found on man-made structures, as well as in pine woods. Length of female 13 to 25 mm; of male 5.4 to 7.9 mm.

Michigan west to the Rockies and to the Pacific Coast States. In Canada from Saskatchewan west to British Columbia.

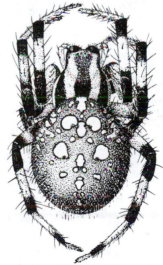

Figure 393. *Araneus trifolium,* with legs held in position assumed while crouching in the retreat.

Figure 393. *Araneus trifolium* (Hentz). Shamrock Spider.

The carapace is white with a central black stripe and a marginal one each side. The legs are conspicuously marked with light and dark annuli. The abdomen may be pale green, or brown to gray, or even purplish red, but the pattern of light spots is always the same, as figured. The male has the abdomen white or yellow, unmarked above, coxae I and II lack spurs, femur II lacks the groove and tibia II is not thickened. Length of female 9 to 20 mm; of male 4.5 to 8 mm.

The snares are large orbs like those of *marmoreus* (fig. 395) but built in tall grass of more open areas.

Throughout the United States but more common in northern than southern States.

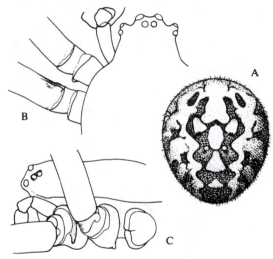

Figure 394. *Araneus marmoreus.* A, Abdomen of female; B, Male from above showing groove on femur II; C, Male from the side to show spur on coxae I and II.

Figure 394. *Araneus marmoreus* Clerck. Marbled Spider.

The cephalothorax is yellow with darker lines along the sides and in the middle. The abdomen is yellow orange with brown to purple markings in a definite pattern as figured. In the male, coxa I has a spur along its posterior edge and femur II has a groove along its anterior edge. Coxa II also has a spur and tibia II is much thickened and curved. Length of female 9 to 18 mm; of male 5.9 to 9 mm.

The webs (fig. 395) are built in wooded areas between trees and shrubs. A signal line connects the hub with a retreat, made by fastening several leaves together somewhere above the snare.

Eastern States to Texas and North Dakota, then across the northern Rockies to Washington and Oregon. Also, all of Canada to Alaska.

Figure 396. *Araneus miniatus* (Walckenaer)

The color is yellow, with paired dark spots on the posterior half of the abdomen, and a broad white transverse band between the shoulder humps. Length of female 3 to 7 mm; of male 2 to 3.7 mm.

New England to Florida and west to Tennessee and Texas.

Araneus partitus (Walckenaer)
Quite similar to *miniatus* but slightly smaller, with the female 3 mm, and the male 2.5 mm.

New England south to Florida and west to Illinois and Arkansas.

Figure 397. *Araneus juniperi* (Emerton)

The abdomen shows a pattern of alternating pale green and white longitudinal bands. Length of female 2.5 to 5.5 mm; of male 2.7 to 4.6 mm.

New England south to Florida and west to Illinois and Texas.

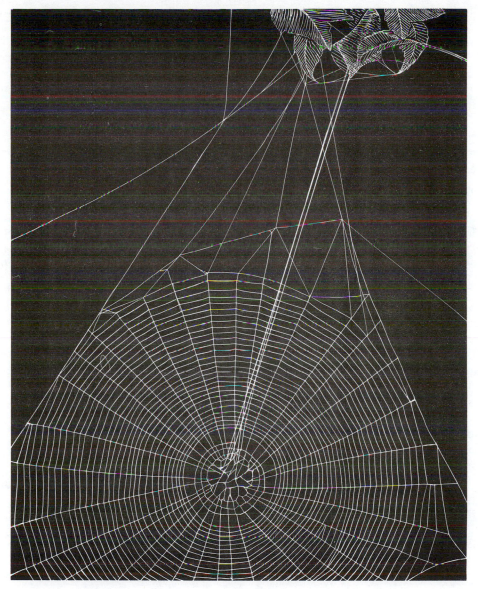

Figure 395. *Araneus marmoreus* snare.

Figure 396. *Araneus miniatus*.

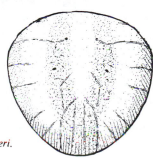

Figure 397. *Araneus juniperi*.

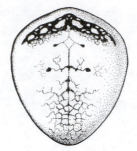

Figure 398. *Araneus niveus,* female.

Figure 398. *Araneus niveus* (Hentz)

The carapace and legs are white to yellow. The abdomen is white on the dorsum and pale green laterally. At the anterior portion is a transverse black band with irregular borders. Some specimens have as well a triangular black mark farther back. In some there are red spots posteriorly. The illustration shows one variety. Length of female 3.2 to 5 mm; of male 2.9 to 4.3 mm.

New Jersey south to Florida and west to Louisiana and Missouri.

Araneus guttulatus (Walckenaer)

Similar to *niveus,* with a large black patch, but with more red on the dorsum, and in general the base color is more greenish. There is much variation in the depth of pigmentation. Length of female 3.8 to 6 mm; of male 3.9 to 4.8 mm.

New England south to Georgia and west to Arkansas and Wisconsin.

Araneus cingulatus (Walckenaer)

Similar to *niveus* and *guttulatus* but lacking the abdominal black patch. In life there may be red and green spots, but these fade to whitish in the preserving fluid. Length of female 4.6 to 6 mm; of male 2.7 to 3.5 mm.

New England south to Florida and west to Texas and Missouri.

Figure 399. *Araneus pratensis.*

Figure 399. *Araneus pratensis* (Emerton)

The carapace is yellow in the middle and light brown toward the sides. The abdomen has two broad longitudinal stripes, brown in front, darkening to almost black behind. Length of female 3.8 to 5.6 mm; of male 3.2 to 4 mm.

In tall grass and low trees.

Eastern United States west to Iowa and Texas.

FAMILY TETRAGNATHIDAE

As far as known the webs are complete orbs with open hubs, built near water. There are relatively few radii and few spirals. There are four genera in our region.

1a **Endites slightly convergent over labium, and not dilated distally. Lateral eyes contiguous or almost so. Legs without spines. Abdomen oval and usually less than one and half times as long as wide** . . . .
. . . . *Pachygnatha* (8 species)

GENUS PACHYGNATHA
Thick-jawed Orb Weavers

Although immature specimens have been seen on webs, the adults do not appear to make any.

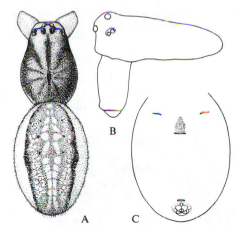

Figure 400. *Pachygnatha tristriata;* A, from above; B, Cephalothorax from the side; C, Venter.

Figure 400. *Pachygnatha tristriata* C.L. Koch

The color is yellowish brown, darker along the sides of the carapace. The abdomen has a folium of silver and gray bordered with black. The posterior medium eyes are not larger than the anterior medians but are elevated so that the carapace is highest at this place. Length of female 5.5 to 6.5 mm; of male 5 to 5.5 mm.

New England and adjacent Canada south to Georgia and west to Texas and Nebraska.

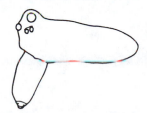

Figure 401. *Pachygnatha autumnalis.*

Figure 401. *Pachygnatha autumnalis* Keyserling

Similar to *tristriata* but with the folium showing a scalloped edge and sometimes with a red stripe in the middle. The posterior median eyes are much larger than the others and are on the sides of a prominence. Length of female 4 to 4.5 mm; of male 3.8 mm.

New England to Georgia and west to Illinois.

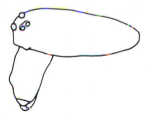

Figure 402. *Pachygnatha xanthostomata.*

Figure 402. *Pachygnatha xanthostomata* C.L. Koch

Similar to *autumnalis,* with the folium scalloped along the edges. The posterior median eyes are not larger than the anterior medians and are not elevated. There is a distinct hump on the chelicera above the fang. Length of female 5.5 to 6 mm; of male 5 to 6 mm.

New England and adjacent Canada to District of Columbia and west to Pennsylvania.

1b **Endites parallel and more or less dilated distally (fig. 403). Lateral eyes of each side not contiguous. Legs spiny. Abdomen long and narrow, usually two or three times as long as wide. . . . .**
. . . . *Tetragnatha* (17 species)

Figure 403. *Tetragnatha* male, cephalothorax from below.

GENUS TETRAGNATHA
Long-jawed Orb Weavers

The two eye rows may be parallel, or diverge (*Arundagnatha*), or they may converge, but the lateral eyes are never contiguous. The chelicerae are well developed, especially in males and the margins of the fang furrow are provided with numerous teeth. In males, in addition to these marginal teeth, there is a strong projecting spur near the base of the fang (fig. 404). The abdomen is quite long and in females often swollen at the base.

Figure 404. *Tetragnatha* male, showing the characteristic projecting spur over the fang.

Most species of *Tetragnatha* build their webs in meadows and in bushes and long grass near water. The snare has few radii and an open hub (fig. 405) at which the spider sits.

Figure 405. *Tetragnatha* snare.

Figure 406. *Tetragnatha laboriosa* female.

Figure 406. *Tetragnatha laboriosa* Hentz

The chelicerae are short and practically vertical, being one-half to two-thirds the length of the carapace in males and shorter in females. The lateral eyes of each side are as far apart as the medians. The general color is light yellow with a silvery abdomen. Length (exclusive of chelicerae) of female 6 mm; of male 5 mm.

Throughout the United States and Canada to Alaska.

Figure 407. *Tetragnatha versicolor,* female.

Figure 407. *Tetragnatha versicolor* Walckenaer

The chelicerae are a little more than half as long as the carapace in females and just a little shorter than the carapace in males. The lateral eyes of each side are closer together than are the medians. The general color is dull yellow brown to gray with a darker broad longitudinal area on the abdomen bordered by a silvery stripe each side. It is never as silvery as *laboriosa,* however. Length (exclusive of chelicerae) of female 6.5 mm; of male 5 mm.

Throughout the United States and southern Canada.

Figure 408. *Tetragnatha straminea,* female.

Figure 408. *Tetragnatha (Arundagnatha) straminea* Emerton

The general appearance is like that of *laboriosa,* but the abdomen is more slender and more yellow than silvery. Because the lateral eyes of each side are farther apart than the medians, this species has been placed by some workers in a separate genus, *Arundagnatha.* Length (exclusive of chelicerae) of female 8 mm; of male 6.5 mm.

New England south to Florida and west to Texas and Minnesota.

Figure 409. *Tetragnatha elongata,* female.

Figure 409. *Tetragnatha elongata* Walckenaer

The abdomen is broad near the base and tapers toward the posterior end. The general color is reddish brown with gray markings on the carapace, and dull silvery with brownish markings on the abdomen. The chelicerae are longer than the carapace in males, and almost as long in females, and the fang is sinuate. The lateral eyes of each side are closer together than are the medians. Length (exclusive of chelicerae) of female 9 mm; of male 7.5 mm.

The egg sacs are attached to twigs and are sparsely covered with strings of beadlike greenish silk (fig. 410).

Throughout the United States, but more common in the East. Also throughout Canada to Alaska.

Figure 410. Egg sac of *Tetragnatha elongata.*

FAMILY AGELENIDAE

Funnel-web Weavers

Most of these spiders spin sheet or platform-like webs with a tube or funnel leading off from the center of one edge. Over the upper side of these webs the spiders run, in an upright position, seize the prey and retire with it to the funnel retreat. There are 25 genera in our region, many of them abundant in the west.

1a With six eyes (fig. 411) the anterior medians lacking[3] . . . . *Yorima* (7 species)

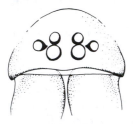

Figure 411. *Yorima,* face.

Figure 412. *Yorima angelica.*

3. *Yorima* is found only in California. Excluded from this book are the members of the genus *Blabomma,* from the Pacific Coast States, which likewise have only six eyes, and some species of *Cicurina,* mainly from Texas caves, which have the eyes reduced to six, or even to zero.

Figure 412. *Yorima angelica* Roth

The carapace and legs are orange. The abdomen is grayish and mottled as illustrated. Length of female 3.3 to 5.5 mm; of male 2.8 to 4.1 mm.

This is a common spider building its web close to the ground in chaparral, and is often caught in pit traps.

California.

1b **With eight eyes** 2

2a **Labium longer than wide. Hind spinnerets with apical segment at least as long as the basal** **5**

2b **Labium as wide as long or wider. Hind spinnerets with apical segment much shorter than the basal** **3**

3a **Posterior spinnerets not longer than anterior (fig. 413). Coxae IV separated by not more than one-sixth their length. Clypeus higher than the diameter of an anterior lateral eye. (Tibia I with two or three pairs of ventral spines.).** ....
.... ***Cybaeus*** **(39 species)**

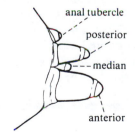

Figure 413. *Cybaeus*, spinnerets.

GENUS CYBAEUS

Virtually all of the species are found in the western States.

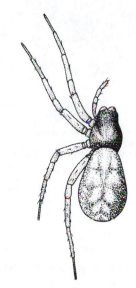

Figure 414. *Cybaeus reticulatus*, female.

Figure 414. *Cybaeus reticulatus* Simon

The carapace is orange brown with dusky markings. The abdomen is yellow with a series of black chevrons. The legs are brown with black annuli. Length of female 9 to 10 mm; of male 7 to 8 mm.

These spiders are most common in wooded areas, especially in mountainous regions. They construct webs under sticks, stones, and debris on the ground. The webs are not as clearly funnel-like as in the other members of the family, nor are the spiders as quick in their movements.

California north to Alaska.

3b **Posterior spinnerets longer than anterior. Coxa IV separated by about one-third to one-half their length (as in fig. 415). Clypeus equal to or less than the diameter of an anterior lateral eye** **4**

Figure 415. *Cryphoeca*, sternum and adjacent structures.

**4a Height of clypeus less than the diameter of an anterior lateral eye (fig. 416). Anterior lateral eyes much larger than the anterior medians. Posterior spinnerets with apical segment not more than one-fifth the length of basal. Retromargin of cheliceral fang furrow with three teeth. . .
. . . . *Cryphoeca* (2 species)**

Figure 416. *Cryphoeca*, face and chelicerae.

Figure 417. *Cryphoeca montana*.

Figure 417. *Cryphoeca montana* Emerton

The carapace is yellow with dusky blotches and a gray marginal stripe. The legs are yellow and the abdomen gray and yellow with a pattern of chevrons. Length of female 3 to 3.5 mm; of male 2.5 to 3 mm.

These spiders live under leaves, stones, and debris on the ground in wooded areas. Superficially they may be mistaken for young specimens of *Callobius bennetti*, the abdominal pattern of the two species being quite similar.

New England west to Wisconsin. Also southern Canada.

4b Height of clypeus about equal to the diameter of an anterior lateral eye (fig. 418). Posterior spimmerets with apical segment hardly half the length of basal segment. Retromargin of cheliceral fang furrow with a series of minute denticles . . . . *Cicurina* (50 species)

Figure 418. *Cicurina*, face and chelicerae.

GENUS CICURINA

Most of the species are western. They all have more or less the same general appearance, with the cephalothorax and legs yellowish orange to brown, and the abdomen white to light gray with darker markings as spots or chevrons. They are generally found under stones and dead leaves on the ground. There are a number of species, mainly from Texas caves, in which the eyes are reduced to six or even zero.

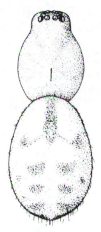

Figure 419. *Cicurina brevis.*

Figure 420. *Agelenopsis*, face and chelicerae.

Figure 421. *Rualena*, tibia IV showing plumose hairs.

Figure 419. *Cicurina brevis* (Emerton)

Length of female 3 to 5 mm; of male 3 to 4 mm.
New England and adjacent Canada south to Georgia and west to the Rockies.

Cicurina arcuata Keyserling
Length of female 4.7 to 7 mm; of male 4.3 to 5.7 mm.
New England and adjacent Canada south to Georgia and west to Louisiana and Missouri.

Cicurina utahana Chamberlin
Length of female 4.2 to 6.2 mm; of male 2.9 to 4.7 mm.
Utah and New Mexico west to California.

Cicurina robusta Simon
The chelicerae are more robust than in the other species. Length of female 5.8 to 9.1 mm; of male 5 to 6.7 mm.
Rocky Mountain and Great Basin States.

5a **Both eye rows very strongly procurved so that the posterior lateral and anterior median eyes from a nearly straight line (fig. 420). Legs and carapace clothed with plumose hairs visible at magnification of about 36X and over (fig. 421) 6**

5b **Eye rows not so strongly procurved. Legs and carapace not clothed with plumose hairs (except for *Tegenaria*) 10**

6a **Retromargin of cheliceral fang furrow with three, rarely four, teeth. Anterior median eyes slightly larger than anterior laterals. Apical segment of posterior spinnerets twice as long as basal segment . . . . *Agelenopsis* (18 species)**

GENUS AGELENOPSIS
Grass Spiders

There are a number of very common species in this genus, all resembling one another closely. They differ principally in genitalia characters, and also somewhat in size and depth of pattern

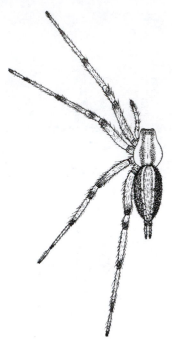

Figure 422. *Agelenopsis,* female.

buildings, and is firmly attached to the substratum. Besides the horizontal sheet with its funnel retreat the spider builds an irregular network or labyrinth which may extend far above the sheet proper. Neither the sheet nor the labyrinth are composed of adhesive threads, but the network serves to impede the flight of insects causing them to fall upon the sheet. The spider depends upon its lightning-like movements to effect a capture.

Figure 423. *Agelenopsis pennsylvanica.*

pigmentation, though even in the same species there is much variation. Generally the carapace is yellowish to brown, with a pair of wide dark bands extending back from the lateral eyes and with a thin dark marginal line each side. The sternum is yellow to brown, often with a V-shaped dark mark. The abdomen is yellowish gray to reddish brown above with a lighter median band on the dorsum. On the venter is a broad gray band whose edges are quite dark so that in many cases there appears to be rather a pair of thin black lines with a light gray area between. The legs are marked with indistinct annuli, which are usually darkest at the distal ends of the segments. The members of this genus are the largest in the family.

The web (fig. 424) is built in grass, on bushes, in stone fences, or in corners of

Figure 423. *Agelenopsis pennsylvanica* (C.L. Koch)

This is one of the grayer species, with a distinct V-shaped mark on the sternum and a broad gray band on the venter. Length of female 10 to 17 mm; of male 9 to 13 mm.

This species is common in open woods and around buildings. The eggs are deposited in the fall. The sacs (fig. 425) are white, shaped like a plano-convex lens about 8 to 9 mm in diameter, and attached by the plane surface to the underside of loose bark. The female remains nearby and dies with the advent of cold weather.

New England south to Tennessee and west to Oregon and Washington.

Figure 424. *Agelenopsis* web.

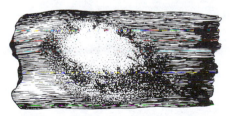

Figure 425. Egg sac of *Agelenopsis*.

Agelenopsis naevia (Walckenaer)

This is the largest and darkest species of the genus. The abdomen is often dark chestnut brown with the markings obscure. The V-shaped mark on the sternum and the broad median band on the venter are usually not distinct, but the annuli on the legs are quite distinct. Length of female 16 to 20 mm; of male 13 to 18 mm.

Found commonly in open fields and among stones of roadside fences. Its webs may reach a large size, up to three feet in width.

New England and adjacent Canada south to Florida and west to Kansas and Texas.

Agelenopsis aperta (Gertsch)
Similar to, and often confused with, *naevia*. Individuals of both sexes vary from about 10 to 19 mm in length, with the males only slightly smaller than the females. Found commonly in the semi-arid regions of the southwest.

Texas to California, and in Colorado and Utah.

6b **Retromargin of cheliceral fang furrow with two (only occasionally with three) teeth. Anterior median eyes not larger than anterior laterals 7**

7a **Anterior median eyes distinctly smaller than anterior laterals . . . .**
. . . . *Rualena* (8 species)

GENUS RUALENA

The members of this genus, all western, closely resemble those of *Agelenopsis* in general appearance.

Rualena cockerelli Chamberlin & Ivie
Length of female 6.5 to 8.5 mm; of male 6 mm.

The spider is usually found in a funnel web in bushes.

California.

7b **Anterior eyes of equal size 8**

8a **Apical segment of posterior spinnerets one and a half to twice as long as basal . .**
. . . . *Calilena* (15 species)

GENUS CALILENA

The members of this genus are all found from Utah west to the Pacific.

Calilena restricta Chamberlin & Ivie
These spiders are usually found in webs on the ground. Length of female 8 to 11.5 mm; of male 6.4 to 9.4 mm.

Arizona north to Idaho.

8b **Apical segment of posterior spinnerets no longer than the basal 9**

9a **Apical segment of posterior spinnerets as long as the basal . . . .**
. . . . *Novalena* (6 species)

GENUS NOVALENA

The members of this genus are found from Colorado west to the Pacific.

Novalena intermedia Chamberlin & Gertsch
This species very closely resembles *Agelenopsis pennsylvanica* in general appearance (fig. 423). Length of female 9 to 10 mm; of male 7.5 to 9 mm.

Pacific Coast States.

9b **Apical segment of posterior spinnerets distinctly shorter than the basal . . . .**
. . . . *Hololena* (30 species)

Figure 426. *Hololena curta*, male.

Figure 426. *Hololena curta* (McCook)

The carapace is orange, with two distinct gray bands laterally. The abdomen is yellowish gray with a broad median light band which encloses a basal gray mark up front. Length of female 10 to 12 mm; of male 9 mm.

    Pacific Coast States.

Hololena hola (Chamberlin & Gertsch)
Similar in appearance to *curta.* Length of female 9 mm; of male 8 mm.

    Colorado, Utah, Arizona and New Mexico.

10a   **Retromargin of cheliceral fang furrow with two teeth. Chelicerae quite strongly geniculate and robust. (fig. 427)** . . . . *Wadotes* (7 species)

Figure 427. *Wadotes,* chelicera and anterior part of carapace, from the side.

GENUS WADOTES

This genus appears limited to the eastern States. The spiders are most often found under stones and loose bark. The carapace is yellowish brown, darkest in front, and the abdomen grayish, with rows of light spots.

Figure 428. *Wadotes hybridus,* female.

Figure 428. *Wadotes hybridus* (Emerton)

Length of female 10 to 14 mm; of male 8.5 to 10.5 mm.

    New England south to Georgia and west to Michigan.

Wadotes calcaratus (Keyserling)
Similar to *hybridus,* but somewhat smaller. Length of female 6 to 11 mm; of male 6 to 8.7 mm.

    New England and adjacent Canada south to Georgia and west to Indiana.

10b   **Retromargin with more than two teeth** **11**

11a   **Chelicerae geniculate and robust (fig. 429); the retromargin of fang furrow with three (rarely four) teeth. Anterior median eyes usually larger than the anterior laterals** . . . . *Coelotes* (5 species) and *Coras* (9 species)

Figure 429. *Coras,* chelicera and anterior part of carapace, from the side.

GENERA CORAS AND COELOTES

The species in these two genera have the same general appearance and are separable on the basis of genitalic structure. The chelicerae, although robust are not as strongly so as in *Wadotes.* The ground color of the carapace is yellowish brown, darkest in front, and with two grayish bands as in the illustration (fig. 430). The abdomen is grayer, with spots.

Figure 430. *Coras medicinalis.*

Figure 430. *Coras medicinalis* (Hentz)

The legs have indistinct gray rings. Length of female 9.5 to 13.5 mm; of male 9 to 13 mm.

Found in cellars of houses, as well as under loose bark, in hollow tree stumps, etc.

Eastern United States and adjacent Canada west to Minnesota and Texas.

Coras lamellosus (Keyserling)
Length of female 8.5 to 13.2 mm; of male 8.5 to 12.5 mm.

Pennsylvania southwest to Mississippi and west to Kansas and Minnesota; but also in Oregon and California.

Coelotes montanus Emerton
Length of female 9.7 to 13.1 mm; of male 8.2 to 9.7 mm.

New England and adjacent Canada west to Minnesota.

Coelotes juvenilis Keyserling
Length of female 6.7 to 11 mm; of male 6.1 to 9.2 mm.

New England south to Virginia and west to Michigan.

11b **Chelicerae not geniculate, less robust; retromargin of fang furrow with 4 to 6 teeth (rarely 3). Anterior median eyes not larger than the laterals, usually smaller 12**

12a **Posterior eye row slightly procurved, with the posterior median eyes very little smaller than the posterior laterals (fig. 431). Legs and carapace clothed with plumose hairs visible at magnifications of 36X and over (fig. 421). Tibia and patella on leg I less than one and a half times the length of the carapace . . . .**
. . . . *Tegenaria* (5 species)

Figure 431. *Tegenaria,* eye region from above.

Figure 432. *Tegenaria domestica,* abdomen.

Figure 432. *Tegenaria domestica* (Clerck)

The carapace is pale yellow with two gray stripes faintly indicated, and the abdomen has a number of irregular gray spots. The legs are long, and faintly annulate. Length of female 7.5 to 11.5 mm; of male 6 to 9 mm.

This species is sometimes taken under stones and in rock crevices, but more often from barns, cellars, and dark corners of rooms. Mature specimens may be found at all seasons, and individuals have been known to live several years. Moreover, this is one of those species in which males and females may live peaceably together on the same web during the two or three months of the breeding season.

Throughout the United States and southern Canada.

12b **Posterior eye row straight or slightly recurved, with the posterior medians much smaller than the posterior laterals (fig. 433). Legs and carapace devoid of plumose hairs. Tibia plus patella of leg I at least one and a half times as long as carapace . . . .**
. . . . *Calymmaria* (10 species)

Figure 433. *Calymmaria cavicola,* eye region from above.

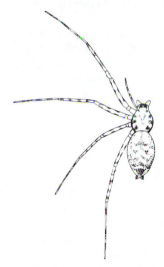

Figure 434. *Calymmaria cavicola,* female.

Figure 434. *Calymmaria cavicola* (Banks)

The carapace and legs are yellowish orange with faint dusky markings. The abdomen is yellow with a faint indication of gray chevrons on the posterior half. The legs are relatively long, and quite thin, and, in the case of the male at least, the femora far exceed the carapace in length. Length of female about 4 to 6 mm; of male about 4 to 5 mm.

Found in caves as well as outside. Virginia south to Florida, west to Alabama, and Indiana.

Calymmaria emertoni (Simon)
Length of female 7 mm; of male 5.5 mm.
Oregon and Washington.

Calymmaria californica (Banks)
Length of female 5 mm; of male 3.8 mm.
California.

FAMILY HAHNIIDAE

These are small spiders which build delicate sheet webs, without retreats, near the ground, usually near water or in moss, rarely in dry places between or under stones. The webs are so delicate that they are generally invisible unless covered with dew. There are three genera in our region.

1a **Anterior median eyes smaller than the anterior laterals. Tracheal spiracle twice as far from the epigastric furrow as from the base of the median spinnerets. Distal segment of lateral spinnerets half as long as the proximal** . . . .
. . . . ***Hahnia*** **(7 species)**

Figure 435. *Hahnia cinerea,* female.

Figure 435. *Hahnia cinerea* Emerton

The abdomen has a double row of oblique light markings on a gray background. The carapace is light to dark brown, with a black pattern, and margined with black. The legs are yellowish brown. Length of female 1.45 to 2 mm; of male about 1.65 mm.

Virtually all of the United States and Canada to Alaska.

1b **Anterior median eyes not smaller than the anterior laterals. Tracheal spiracle**
twice as far from the base of median spinnerets as from the epigastric furrow. Segments of lateral spinnerets subequal (fig. 436) . . . .
. . . . ***Neoantistea*** **(11 species)**

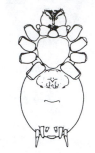

Figure 436. *Neoantistea.*

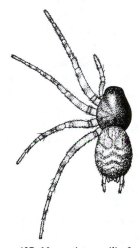

Figure 437. *Neoantistea agilis,* female.

Figure 437. *Neoantistea agilis* (Keyserling)

The abdomen is gray with a pattern of yellow spots. The carapace is reddish or orange brown. The legs are yellowish, ringed with gray. Length of female 2.5 to 3.2 mm; of male 2.25 to 2.6 mm.

Throughout the United States and Canada.

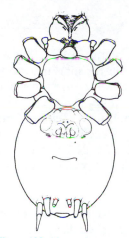

Figure 438. *Neoantistea magna.*

Figure 438. *Neoantistea magna* (Keyserling)

Similar in appearance to *agilis* but larger. Length of female 3.4 to 3.6 mm; of male 2.7 to 3.2 mm.

New England south to Florida and west through the northern States to northern California. Also all Canada to Alaska.

FAMILY MIMETIDAE

Pirate Spiders

The characteristic spination of the legs (fig. 439) enables one readily to distinguish the two genera of this family from all others. The chelicerae are fairly long and slender and, at least in our species, are fused at the base. They are without a boss and provided on the promargin of the fang furrow with a series of long appressed bristles.

These spiders are not known to build snares and are predatory on other spiders. In fact, from some accounts they seem not to eat insects at all but restrict themselves to spiders though some have been reported eating insects.

Besides being found on the webs of other spiders they have been taken from bushes, and from under loose stones on the ground.

Figure 439. *Mimetus*, metatarsus I showing spination.

1a Height of clypeus from one-third to one-half that of median ocular area. Leg I one and half times as long as leg IV. Chelicera with a conspicuous heavy bristle on the inner margin about two-thirds the distance from base to fang-groove . . . . *Mimetus* (10 species)

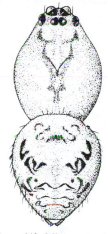

Figure 440. *Mimetus puritanus.*

Figure 440. *Mimetus puritanus* Chamberlin

The general color is yellow with dark blotches along the middle of the carapace and on the abdomen. There are dark spots on the ventral surface of femora I and II, and two on the front of each chelicera. Length of female 5 to 5.6 mm; of male 4 to 4.5 mm.

Taken from bushes, New England and adjacent Canada south to Georgia and west to Wisconsin and Kansas.

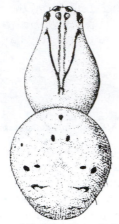

Figure 441. *Mimetus epeiroides.*

Figure 441. *Mimetus epeiroides* Emerton

The general color is yellow with four thin black lines extending from the eyes to converge at the dorsal furrow. There are black spots on the abdomen, two thin black lines on the ventral surface of femora I and II, and a single black spot on the front of each chelicera near its base. Length of female 4.4 to 4.7 mm; of male 3.4 mm.

New England west to Utah.

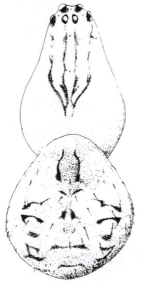

Figure 442. *Mimetus hesperus.*

Figure 442. *Mimetus hesperus* Chamberlin

The pattern on the carapace is similar to that of *epeiroides,* but the markings on the abdominal dorsum are more like those of *puritanus,* though not quite. Also, as in *puritanus* there are black spots under femora I and II, and two on the front of each chelicera. Length of female 4 to 6.3 mm; of male 3.5 to 4.5 mm.

Utah and Texas west to the Pacific Coast States.

1b Clypeus hardly narrower than the length of median ocular area. Leg I not more than one and a third times as long as leg IV. Chelicera without heavy bristle . . . . .
. . . . . *Ero* (4 species)

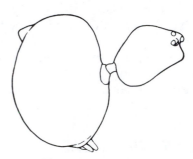

Figure 443. *Ero leonina.*

Figure 443. *Ero leonina* (Hentz)

The general color is light yellow to pale gray with markings of brown to dark gray. Length of female 2.7 to 3.4 mm; of male 2.3 to 2.6 mm. The cephalothorax is quite high in the middle and slopes steeply to the rear. The abdomen has a pair of conical tubercles on the highest part.

Found under stones and dead leaves on the ground, as well as in grass and low bushes. The egg sacs (fig. 444) are roughly spherical, about 3.5 mm in diameter, of loose texture and pale brown color, enclosed in an irregular network of coarse dark reddish threads and suspended by a cord of coarse threads 5 to 25 mm long.

Throughout the United States and southern Canada.

Figure 444. Egg sac of *Ero*.

FAMILY PISAURIDAE

Nursery-web Spiders

These are somewhat similar to the wolf spiders but show a smaller difference in size between the eyes and have the head region lower. Except in some juveniles a snare is not built, the prey being hunted as in the Lycosidae. The egg cocoon is spherical, of two halves, and is carried by the female under the sternum, held in place by chelicerae and pedipalps (fig. 445). Just before the spiderlings emerge the mother fastens the sac to some leaves, builds a nursery web around it and mounts guard nearby (fig. 446). There are five genera in our region.

Figure 445. *Pisaurina* carrying her egg sac.

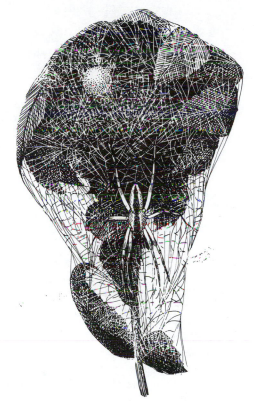

Figure 446. *Pisaurina* standing guard below her nursery.

1a **Anterior row of eyes strongly procurved and posterior eye row strongly recurved so that there appear to be four rows of two each, the anterior laterals constituting the front row and not more than a diameter from the edge of the clypeus (fig. 447)** . . . . *Pelopatis* (1 species)

Figure 447. *Pelopatis*, face.

Figure 448. *Pelopatis undulata,* female.

Figure 448. *Pelopatis undulata* (Keyserling)

The cephalothorax is very much longer than broad, which together with the elongate abdomen and the yellow color of body and legs makes this spider resemble a *Tibellus* (Family Philodromidae). There is a thin black line running along the middle of the carapace, and there are three teeth on the retromargin of the cheliceral fang furrow. The tibiae I and II are provided with five pairs of ventral spines, which, except for the apical pair, are very long and overlapping. The metatarsi I and II are provided with four pairs of very long ventral spines. The sternum is longer than wide and produced at the posterior end to extend between coxae IV. Length of female 11 to 19 mm; of male 9 to 14 mm.

North Carolina south to Florida and west to Louisiana. Virtually always found close to the edge of ponds and streams.

1b   **Anterior row of eyes straight or only slightly curved, so that there are quite definitely four eyes in the front row, with the anterior lateral eyes at least two diameters from the edge of the clypeus. . 2**

2a   **Height of clypeus equal to the length of the median ocular area. Retromargin of cheliceral fang furrow with four teeth. Lorum of pedicel with the anterior sclerite (a) notched behind (c) to receive a projection of the posterior (fig. 449) . . . .**
. . . . *Dolomedes* (9 species)

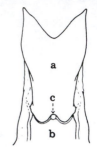

Figure 449. *Dolomedes,* lorum.

GENUS DOLOMEDES
Fishing Spiders

Most of the members of this genus live near water and have been reported catching small fishes as well as the aquatic insects on which they usually feed.

Figure 450. *Dolomedes scriptus,* male.

Figure 450. *Dolomedes scriptus* Hentz

The colors are brownish gray, with white areas more extensive than in *tenebrosus,* which it resembles. Length of female 17 to 24 mm; of male 13 to 16 mm.

New England and adjacent Canada south to Florida and west to South Dakota and Texas.

Figure 451. *Dolomedes tenebrosus,* female.

Figure 451. *Dolomedes tenebrosus* Hentz

The colors are brownish gray with less white than in *scriptus,* which it resembles. The white W-shaped marks on the abdomen are lacking. Length of female 15 to 26 mm; of male 7 to 13 mm.

This species is often found some distance from water, in wooded areas.

New England and adjacent Canada south to Florida and west to North Dakota and Texas.

Figure 452. *Dolomedes triton,* male.

Figure 452. *Dolomedes triton* (Walckenaer)

This species is also known under the name *sexpunctatus* Hentz. The ground color is greenish brown with a pattern of contrasting white lines and spots. The sternum is marked with six large black spots. The male has a spinose hump on ths ventral side of femur IV. Length of female 17 to 20 mm; of male 9 to 13 mm.

This species is found running along the edge of ponds, or over floating vegetation.

Throughout the United States and southern Canada, although seldom encountered in the Rockies and Great Plains.

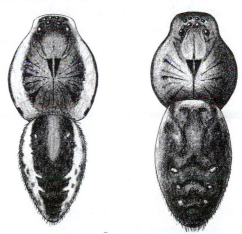

Figure 453. *Dolomedes vittatus,* A, male; B, female.

Figure 453. *Dolomedes vittatus* Walckenaer

The pattern on the carapace is characteristic, with a pair of triangular dark spots in front of the thoracic groove. As in *triton* the male has a spinose hump on the ventral side of femur IV. Length of female 19 to 28 mm; of male about the same.

New England and adjacent Canada south to Florida and west to Texas and Oklahoma.

2b Height of clypeus less than the length of the median ocular area. Retromargin of cheliceral fang furrow with only three teeth. Lorum with transverse, or slightly procurved suture (c) between the two sclerites (fig. 454)..................3

Figure 454. *Pisaurina,* lorum of pedicel.

3a Median ocular area much wider behind than in front and wider than long. Of the retromarginal teeth the two outer are close together and are removed from the inner tooth by a space (a) as wide as the latter (fig. 455) ....
...................*Tinus* (1 species)

Figure 455. *Tinus,* right chelicera from below showing teeth.

Figure 456. *Tinus peregrinus.*

Figure 456. *Tinus peregrinus* (Bishop)

The carapace is yellow to light brown, with a broad gray median longitudinal band and dusky marginal stripes. The abdomen is provided with a dark brown central band margined with white lines. The sternum is yellow with three pairs of light gray spots. Length of female 10 mm; of male 9 mm.

This is the northernmost representative of a genus most of whose members are tropical. It closely resembles a *Dolomedes* superficially.

Texas and Arkansas west to California.

3b Median ocular area as long as wide behind, where it is only slightly wider than in front (fig. 457). Retromarginal teeth evenly spaced ....
...............*Pisaurina* (4 species)

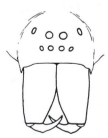

Figure 457. *Pisaurina,* face.

Figure 458. *Pisaurina mira.*

Figure 458. *Pisaurina mira* (Walckenaer)

The ground color is yellow to light brown with a darker brown broad medium band which is margined with white on the abdomen. Another variety has the broad band less distinct. Length of female 12.5 to 16.5 mm; of male 10.5 to 15 mm.

Found in tall grass and on bushes. See also Figs. 445 and 446.

New England and adjacent Canada south to Florida, west to Texas, and Kansas.

FAMILY LYCOSIDAE

Wolf Spiders

We have 13 genera in our region. In this family the eyes are of unequal size, all dark in color, and with the posterior row so strongly curved as to form two rows of two each. The anterior eyes are the smallest and the posterior medians (i.e., the second row) the largest by far. The chelicerae are relatively strong, with toothed margins and with prominent boss. The legs are usually scopulate and spinose.

Except in *Sosippus* and the juveniles of some others, snares are not built. Some make tubular tunnels in the ground, some make use of natural depressions under rocks, etc., while others never use a retreat, but are found running through grass, among dead leaves on the forest floor, or over sandy and stony areas, etc. Many wolf spiders are active at night, as well as during daylight hours, and may be collected at night with the use of a headlight.

The egg sac is globular and of two halves with a seam around its "equator." It is carried by the female attached to her spinnerets (fig. 459). After emergence the young climb up onto the mother's abdomen and are carried about by her for a considerable time (fig. 460).

Figure 459. *Lycosa* carrying her egg sac.

1a **Posterior spinnerets distinctly longer than the anterior, with the apical segment conical and at least half as long as the basal. Retromargin of cheliceral fang furrow with usually four stout teeth, sometimes five. (Anterior row of eyes longer than the second row, and females without spines above on tibiae III and IV) . . . .**
. . . . *Sosippus* (4 species)

GENUS SOSIPPUS

In habits these spiders are atypical for the family. They build sheet webs, with a funnel retreat, over which they run like members of the Agelenidae.

Figure 460. Lycosid female with her spiderlings.

Figure 461. *Sosippus mimus.*

Figure 461. *Sosippus mimus* Chamberlin

The carapace and abdominal dorsum show white markings on a gray or dark brown ground. Length of female 12.9 to 18.2 mm; of male 13.1 to 14.2 mm.

    Florida west to Louisiana.

Sosippus floridanus Simon
The carapace shows a light median band extending back from the second row of eyes, and a white submarginal band close to each lateral edge. The abdominal dorsum is similar to that in *mimus.* Length of female 11.6 to 14.9 mm; of male 11.2 to 11.9 mm.

    Florida.

Sosippus californicus Simon
Similar to *floridanus,* but the cardiac area has the dark pigment more heavily filled in. Length of female 13.7 to 18.6 mm; of male 12.8 to 15 mm.

    Southern Arizona and southern California.

1b Posterior spinnerets at most only slightly longer than the anterior and apical segment hemispherical and very short. Retromargin of cheliceral fang furrow usually with not more than three teeth . . 2

2a Anterior eye row decidedly procurved, the anterior medians much farther from the anterior laterals than from each other. Head wider at top than at clypeus (fig. 462) and with both the posterior medians and posterior lateral eyes at the edges of the head. . . . .
. . . . *Trabea* (1 species)

Figure 462. *Trabea,* face.

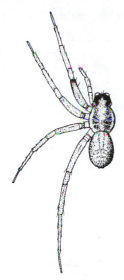

Figure 463. *Trabea aurantiaca,* female.

Figure 463. *Trabea aurantiaca* (Emerton)

The carapace has a yellow marginal stripe each side and a large yellow spot in the middle just behind the eyes. Elsewhere it is black. The abdomen is orange with yellow spots. The front legs are dark brown and the hind legs yellow, with the second and third legs intermediate in color. Length of female 3 to 3.5 mm; of male 2.5 to 3 mm.

Found among dead leaves in damp woods, in swamps, and along the edge of streams.

New England south to Florida and west to Wisconsin.

2b **Anterior eye row not so procurved, the anterior medians hardly, if at all, farther from the anterior laterals than from each other. Head either with the sides sloping so as to be narrower at the top than at the clypeus, or with the sides virtually vertical** 3

3a **Carapace as high in the thoracic region as the cephalic (fig. 464); cephalic region with a dark V-shaped mark within a central pale area. Distal pair of ventral spines on tibiae I and II never in the apical position. Tibia I below with at most three pairs of spines** . . . .
. . . . *Pirata* **(24 species)**

Figure 464. *Pirata,* carapace from the side.

GENUS PIRATA

The members of this genus are usually easily recognized by the characteristic pattern on the carapace. There is a light yellow band extending from the eye region back to the posterior edge, somewhat narrowed behind. Enclosed in this light is a dark V-shaped mark extending from between the rear eyes to the dorsal groove. On either side of the median light area the carapace is dark brown, gray, or black, to a marginal or submarginal light band. The abdomen usually has a light longitudinal stripe on the anterior half and often indistinct chevrons behind. Generally there are paired yellow spots, or spots of white scales on the posterior half. The thoracic part is as high as the cephalic or even slightly higher.

These spiders occur in damp situations as along the margins of ponds and streams. The egg sac is spherical and white.

Figure 465. *Pirata minutus.*

Figure 465. *Pirata minutus* Emerton

The carapace is provided with a thin marginal stripe of white scales. The abdomen is mostly gray with a median basal stripe, often indistinct, and has paired spots of white hairs on the posterior half. Some of the pairs of spots appear joined by light transverse lines. Length of female 3.3 to 3.7 mm; of male 2.7 to 3 mm.

New England and adjacent Canada south to Florida and west to Wisconsin.

Figure 466. *Pirata piratica,* female.

Figure 466. *Pirata piratica* (Clerck)

This species lacks the thin marginal black stripe on the carapace, and instead has a broad yellow band. The abdomen shows a narrow lanceolate basal mark as illustrated. There are numerous white hairs sparsely distributed over the gray dorsum, these being somewhat more abundant on the sides. The femora are yellow. Length of female 5.8 to 7.5 mm; of male 5.5 to 6.2 mm.

New England southwest to West Virginia and across the northern half of the country then southwest to Arizona and the Pacific Coast States. Also, virtually all of Canada to Alaska.

Figure 467. *Pirata montanus.*

Figure 467. *Pirata montanus* Emerton

The abdomen is reddish brown with a gray-bordered basal stripe, and black chevrons on the posterior half. At the lateral ends of each chevron is a spot of white hairs. There is a thin marginal white stripe on the carapace. Length of female 4.7 to 5.3 mm; of male 4.3 to 5.3 mm.

New England west to Utah.

Figure 468. *Pirata insularis*.

Figure 468. *Pirata insularis* Emerton

The markings are quite distinct as figured. Length of female 4 to 5.3 mm; of male 4.5 to 5 mm.

New England and adjacent Canada south to Florida, and west to Oklahoma and Nebraska.

Figure 469. *Pirata maculatus*.

Figure 469. *Pirata maculatus* Emerton

This species closely resembles *insularis* but the submarginal yellow bands on the carapace are not so sharply set off from the gray bands as in that species. Length of female 4.4 to 5 mm; of male 4.5 mm.

Eastern states west to Kansas.

3b **Carapace usually higher in the head region (fig. 470); dark V-shaped mark lacking. Distal pair of ventral spines on tibiae I and II apical in position** **4**

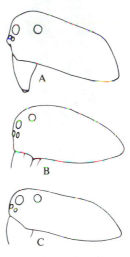

Figure 470. Carapace from the side. A, *Pardosa;* B, *Geolycosa;* C, *Lycosa*.

4a **Retromargin of cheliceral fang furrow with two teeth. (Carapace with a distinct median longitudinal pale band)** . . . . *Tarentula* **(7 species)**

GENUS TARENTULA

It is unfortunate that this generic name is so similar to *Tarantula,* the generic name for an amblypygid whip-scorpion, and to "tarantula," the common name for an entirely different kind of spider. On this account some prefer to use the name *Alopecosa* for spiders of this genus.

Figure 471. *Tarentula aculeata,* male.

Figure 471. *Tarentula aculeata* (Clerck)

The general color is brown with lighter markings. Length of female 11 to 12 mm; of male 8 to 9 mm.

These spiders make shallow holes under stones and are found more commonly in the colder regions of the continent, as on mountains.

New England and adjacent Canada, New York and the Rocky Mountains into Alberta.

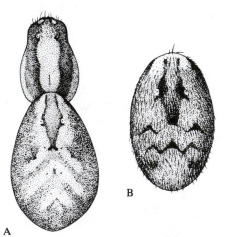

Figure 472A. *Tarentula kochii,* female.
Figure 472B. *Tarentula kochii,* male abdomen.

Figures 472A and 472B. *Tarentula kochii* Keyserling

The general color is brownish gray, with lighter markings as illustrated. The tan central band is sparsely covered with white hairs, and is pinched in at the sides at about the level of coxae II. The abdominal dorsum has a somewhat lighter median area, the pattern being much more distinct in males. Length of female 9 to 13.5 mm; of male 7.5 to 9.2 mm.

Montana and adjacent Canada to New Mexico west to Oregon and California. This is one of the commonest spiders taken in pit traps in southern California.

4b   Retromargin of cheliceral fang furrow with three teeth **5**

5a   Cephalothorax very high in front, and usually sloping in an unbroken line to the posterior edge (see fig. 470B). Females lacking dorsal spines on tibia IV . . . .
. . . . *Geolycosa* **(16 species)**

GENUS GEOLYCOSA
Burrowing Wolf Spiders

Almost their whole existence is spent in the burrow, at the mouth of which they await their prey (fig. 473). They are quite sensitive to the slightest vibrations of the ground and scurry beneath at the first sign of danger. To observe one it is necessary to remain motionless near the burrow for several minutes until the spider returns to the surface. Correlated with the digging habit the front legs are heavy and strong.

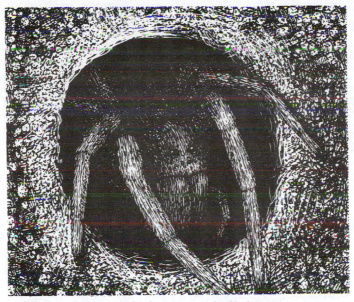

Figure 473. *Geolycosa* at mouth of its burrow.

Figure 474. *Geolycosa pikei.*

Figure 474. *Geolycosa pikei* (Marx)

The general color is sandy gray to brown, with black markings. The males ars lighter gray so that there is much more contrast, and the underside of leg I is jet black. Length of female 18 to 22 mm; of male about 14 mm.

In general, this species builds its burrows in sandy areas near the seashore, though they have also been found inland. Usually no turret is built around the mouth of the tunnel, which may be 10 to 12 inches deep and five-eighths of an inch in diameter. The burrows are not closed during the winter and the spiders remain dormant at the bottom.

New England south to Georgia.

Geolycosa turricola (Treat)

Similar to *pikei,* but with the underside of coxa and femur I not black. Length of female 22 mm; of male 21 mm.

This species usually constructs a turret around the mouth of its burrow, dug in grasslands and open fields.

New England south to Florida and west to Pennsylvania and Tennessee.

Geolycosa missouriensis (Banks)

In general appearance this resembles the preceding two species. However, in addition to having the coxa and femur I light below the patella I is also light below. Also, the venter is light, and dark in the other two. Length of female 21 mm; of male 15 to 18 mm.

New York west to the Mississippi Valley and Great Plains States and adjacent Canada, south to Texas and west to Arizona and Utah.

5b Tibia IV with at least one, usually two or three spines above. Cephalothorax not so high in front, and with more or less of a declivity in the thoracic region (fig. 470C) 6

6a Tibia IV above with the proximal spine usually thinner or more drawn out than the distal one; sometimes reduced to a bristle (figs. 475 and 476) 7

Figure 475. *Trochosa,* tibia IV from above.

Figure 476. *Arctosa,* tibia IV from above.

6b Tibia IV with the two dorsal spines about equally stout (fig. 477) 9

Figure 477. *Lycosa,* tibia IV from above.

7a Carapace hirsute . . . .
. . . . *Trochosa* (6 species)

Figure 478 and 492A. *Trochosa terricola* (Thorell)

This species has long been known under the name *pratensis* (Emerton). The colors are yellow and brown with a lighter, narrow, submarginal stripe on each side of the carapace and a broad median band. Length of female 9 to 14 mm; of male 9 to 12 mm.

These spiders are found running over dead leaves on forest floors, and under stones in fields.

New England and adjacent Canada to North Carolina and west to Washington.

7b Carapace glabrous or nearly so 8

8a Tarsus I with a dorsobasal bristle which is drawn out thin and fine at the end, and is much longer than the hairs and trichobothria (as in fig. 479) . . . .
. . . . *Arctosa* (12 species)

Figure 479. *Arctosa,* tarsus I to show dorsobasal bristle.

Figure 478. *Trochosa terricola.*

Figure 480. *Arctosa rubicunda.*

Figure 480. *Arctosa rubicunda* (Keyserling)

The cephalothorax and legs are evenly dark brown without markings. This species belongs to the section of the genus in which the carapace is more glabrous and the anterior row of eyes is wider than the second row. Length of female 9 to 12 mm; of male 7 to 8 mm.

Found under stones in pastures and under logs in wooded areas.

New England south to District of Columbia and west to Nebraska. Also southern Canada.

Figure 481. *Arctosa littoralis.*

Figure 481. *Arctosa littoralis* (Hentz)

This species is a spotted gray or dirty white, like the color of the sand over which it runs so that the spider is difficult to see when it is not moving. It belongs to that section of the genus in which the carapace is not so glabrous and the anterior row of eyes is shorter than the second. Length of female 11 to 15 mm; of male the same.

Throughout the United States and eastern Canada.

8b Tarsus I without such a bristle . . . .
. . . . *Allocosa* (7 species)

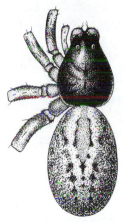

Figure 482. *Allocosa funerea.*

Figure 482. *Allocosa funerea* (Hentz)

The carapace is very dark and shining and at least the first two femora are black with the other segments light in color. Length of female 5 to 6 mm; of male 3.5 to 4.5 mm.

New England to Georgia and west to the Mississippi.

9a Labium not longer than wide, usually wider than long, with the basal articular notches (a) about one-fourth its length (fig. 483). Sides of face vertical or almost so (fig. 484). Metatarsus IV usually longer than tibia plus patella IV, or at least not shorter . . . .
. . . . *Pardosa* (80 species)

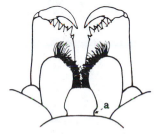

Figure 483. *Pardosa,* mouthparts from below.

Figure 484. *Pardosa*, face.

GENUS PARDOSA
Thin-legged Wolf Spiders

The legs ars relatively long, with the metatarsi and tarsi quite thin, and with very long spines. The tibia plus patella IV is usually longer than the carapace, and tibia I is provided with three pairs of spines of which the distal pair is apical and shorter than the others, which are very much longer than the thickness of the segment. The cephalothorax is highest in the head region and the chelicerae are much smaller than in most other lycosids, so that their height is less than the height of the head (see figs. 470A and 484).

The cocoon is lenticular, usually greenish when fresh, changing to a dirty gray when older.

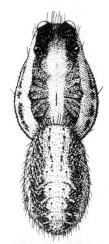

Figure 485. *Pardosa distincta*.

Figure 485. *Pardosa distincta* (Blackwall)

There is a distinct pattern of light brown markings on a yellow background. Length of female 4.7 to 6.3 mm; of male 4 to 4.7 mm.

This species is abundant in grassy fields.

New England and adjacent Canada south to North Carolina and west to Arizona and Montana.

Pardosa utahensis Chamberlin
Similar to *distincta* but somewhat darker yellow, and with the median band on the abdominal dorsum darker than that on the carapace. Length of female 5.2 mm; of male 4.8 mm.

Wyoming, Colorado and Utah at high elevations.

Figure 486. *Pardosa milvina*.

Figure 486. *Pardosa milvina* (Hentz)

This species and *saxatilis* are similar but the yellow spots on the abdomen are larger and more distinct in this than in *saxatilis*. The males can be distinguished easily by the white hairs on the palpal patella of *saxatilis*. Length of female 5.2 to 6.2 mm; of male 4 to 4.7 mm.

Found in dry open woods, as well as on wet ground and along the edges of ponds and streams.

New England and adjacent Canada south to Florida and west to the Rockies.

Figure 487. *Pardosa saxatilis.*

Figure 487. *Pardosa saxatilis* (Hentz)

This species and *milvina* are similar, with some of the differences pointed out under that species. Length of female 4.7 to 5.7 mm; of male 3.8 to 4.5 mm.

Found commonly in grassy fields.

New England and adjacent Canada south to Florida and west to the Rockies.

Figure 488. *Pardosa ramulosa.*

Figure 488. *Pardosa ramulosa* (McCook)

This species is very similar in pattern to *milvina,* and is also almost identical in general appearance to *sternalis,* with which it has been confused. The yellow spots on the abdomen are quite distinct. In males the carapace is much darker so that the pattern is not as clear. Length of female 5 to 7 mm; of male 4.6 to 5.5 mm.

Utah, Nevada and California, being especially abundant in the southern part of California.

Pardosa sternalis (Thorell)
This species is much like the preceding, *ramulosa,* with which it has been confused. Length of female 6 to 7 mm; of male 5.5 to 6 mm.

Abundant in the Great Plains States, and extending across the Rockies and Alberta to Oregon and California.

Pardosa altamontis Chamberlin & Ivie
Similar to *sternalis* and about the same size.

Montana and Utah west to Oregon and California at high elevations.

Figure 489. *Pardosa lapidicina.*

Figure 489. *Pardosa lapidicina* Emerton

The ground color is generally gray to black with most specimens showing yellow spots on the abdomen, and not so distinct lighter areas on the carapace, as figured. Length of female 7.7 to 9.3 mm; of male 6 to 7 mm.

These spiders run about swiftly over rocky shores, among stones and clay banks, and among rocks on talus slopes.

New England south to North Carolina and west to Texas and Nebraska.

Pardosa sierra Banks
Similar to *lapidicina*. Length of female 6.3 to 9.3 mm; of male 5.8 to 7.7 mm.
Wyoming west to Oregon and south to Arizona and California.

Pardosa steva Lowrie & Gertsch
Similar to *lapidicina*. Length of female 6.7 to 8.8 mm; of male 5.8 to 8.4 mm.
Rocky Mountain States north to Alberta and west to Oregon and California.

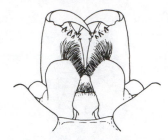

Figure 491. *Lycosa,* mouthparts from below.

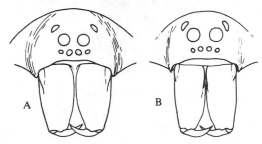

Figure 492. Face and chelicerae. A, *Trochosa;* B, *Lycosa.*

GENUS SCHIZOCOSA

The members of this genus resemble *Lycosa* and are properly distinguished by characters of the genitalia. However, the males of some species are easily recognized by the brushes of hairs on the anterior legs. The spiders do not dig holes in the ground but are found running about over the surface.

Figure 490. *Pardosa xerampelina.*

Figure 490. *Pardosa xerampelina* (Keyserling)

The colors are mainly brownish gray with lighter areas as figured. Length of female 7.4 to 9.5 mm; of male 5 to 6 mm.
New England and adjacent Canada west to Washington.

9b   Labium longer than wide with the basal articular notches usually about one-third its length (fig. 491). Sides of face slanting so that head at top is narrower than at clypeus (fig. 92). Metatarsus IV shorter than tibia plus patella IV, or at least not longer . . . . *Schizocosa* **(17 species) and** *Lycosa* **(45 species)**

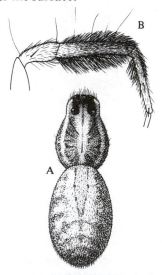

Figure 493. *Schizocosa bilineata.* A, Female; B, Leg I of male.

Figure 493. *Schizocosa bilineata* (Emerton)

The colors are yellow and light brown. In the male, tibia I and the proximal half of metatarsus I have a very conspicuous brush of black hairs. The legs are otherwise yellow without annuli. Length of female 7 to 8 mm; of male 5 to 6 mm.

Found in tall grass.

New England south to Georgia and west to Kansas.

Figure 495. *Schizocosa avida.*

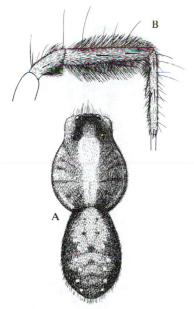

Figure 494. *Schizocosa crassipes.* A, Female; B, Leg I of male.

Figure 494. *Schizocosa crassipes* (Walckenaer)

The general color resembles dead leaves, being dark brown with lighter brown stripes and spots. On the male there is a conspicuous brush of black hairs on tibia I, a smaller brush on the patella, and hardly any on the metatarsus. Length of female 6 to 8 mm; of male the same.

Found running over forest floor litter.

New England south to Georgia and west to Oklahoma and Nebraska.

Figure 495. *Schizocosa avida* (Walckenaer)

The markings of this species are variable, but one of the commonest patterns is illustrated here. The colors are brown and gray. Length of female 10 to 15 mm; of male 8 to 11 mm.

Throughout the United States and southern Canada.

GENUS LYCOSA

This is a very large genus including most of the larger-bodied wolf spiders. Some species build retreats, consisting of either a shallow excavation under a stone, or a tube running vertically or diagonally into the ground. Others are wanderers all the time. The cocoon is spherical, usually white (sometimes green) at first, but changing with age, as it is carried, to dirty gray or brown.

Figure 496. *Lycosa carolinensis*.

Figure 496. *Lycosa carolinensis* Walckenaer

The carapace is dark brown with gray hairs (lighter in males) and usually without distinct markings. The abdomen is likewise brown with a somewhat darker median longitudinal stripe. The sternum, coxae and venter are all black. Length of female 22 to 35 mm; of male 18 to 20 mm. This species usually builds a burrow in the ground.

New England south to Florida and west
Throughout the United States.

Lycosa aspersa Hentz

Similar in appearance to *carolinensis* although the males are somewhat browner. Also, there is a distinct narrow line of yellow hairs in the eye region, lacking in *carolinensis*. Moreover, the venter is not evenly black, but more spotted, and the legs are annulate. This species usually builds a burrow surmounted by a turret. Length of female 18 to 25 mm; of male 16 to 18 mm.

New England and adjacent Canada south to Florida and west to Nebraska.

Figure 497. *Lycosa punctulata*.

Figure 497. *Lycosa punctulata* Hentz

The ground color is yellow, with brownish to black longitudinal stripes. The venter is spotted. Length of female 11 to 17 mm; of male 13 to 15 mm.

New England south to Florida and west to the Rockies.

Figure 498. *Lycosa rabida,* abdomen.

Figure 498. *Lycosa rabida* Walckenaer

This species is similar in pattern and colors to the preceding but the median dark band of the abdomen is broken and encloses lighter areas. In addition, the male has leg I dark brown or black, and the venter is not spotted. Length of female 16 to 21 mm; of male 11 to 12 mm.

New England to Florida and west to Oklahoma and Nebraska.

Figure 499. *Lycosa helluo.*

Figure 499. *Lycosa helluo* Walckenaer

The general color is dull yellow to greenish brown. The venter is spotted. Length of female 18 to 21 mm; of male 10 to 12 mm.

New England and adjacent Canada south to Florida and west to the Rockies.

Figure 500. *Lycosa frondicola,* male.

Figure 500. *Lycosa frondicola* Emerton

Like *gulosa* the general color is dark brown with light gray markings. The venter is black. Length of female 11 to 14 mm; of male 9 to 12 mm.

Found running over dead leaves on the forest floor.

New England and adjacent Canada south to North Carolina and west to the Rockies.

Figure 501. *Lycosa antelucana.*

Figure 501. *Lycosa antelucana* Montgomery

The general color is orange brown with some lighter yellowish areas. On the carapace is a white line between the median eyes. This line becomes wider farther back, much so just before the thoracic groove, where it becomes yellowish, and then it narrows again. On the abdominal dorsum is a dark brown cardiac mark. Length of female 13.5 to 19 mm; of male 13 to 18 mm.

Florida, Alabama and Tennessee west across the southwestern States to California.

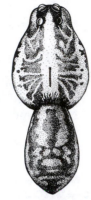

Figure 502. *Lycosa lenta,* male.

Figure 502. *Lycosa lenta* Hentz

The pattern is more distinct in the male than in the female. The colors are brown and yellow above. The sternum and coxae are dark brown to black. The venter behind the epigastric furrow is black, as is also the middle of the epigastrium. The patellae are pale beneath and the femora are unmarked. Length of female 17 to 22 mm; of male 13 to 20.5 mm.

Pennsylvania south to Florida and west to Ohio and Texas.

Lycosa baltimoriana (Keyserling)
In general similar to *lenta,* but with the markings somewhat more pronounced. However, the coxae are not black beneath, though the sternum may be. The venter is black only posterior to the epigastric furrow. The patellae are black beneath, and femora I and II have a dark line on the retrolateral surface. Length of female 15 to 18 mm; of male 14 to 17.5 mm.

New England southwest to Colorado.

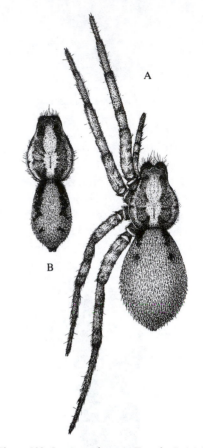

Figure 503. *Lycosa gulosa.* A, Female; B, Male.

Figure 503. *Lycosa gulosa* Walckenaer

The general color is dark brown with light gray markings on the carapace, and black on the abdomen. In the male the black is more extensive. Length of female 11 to 14 mm; of male 10 to 11 mm.

Found running over dead leaves on the forest floor.

New England and adjacent Canada south to Georgia and west to the Rockies.

FAMILY OXYOPIDAE

Lynx Spiders

There are three genera in our region. These spiders are easily recognized by the arrangement of the eyes. The legs are conspicuously spinose (fig. 504) the spines standing out at a considerable angle. The abdomen often tapers to a point behind. The spiders run rapidly and can also jump. They build no snares (except for the juveniles of some), or retreats, or molting nests, and live among low bushes and herbaceous vegetation to which they fasten their egg sacs and over which they hunt their prey.

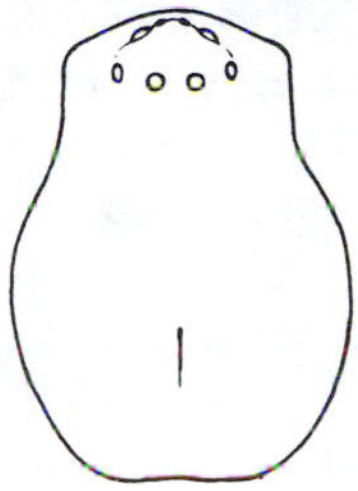

Figure 505. *Peucetia,* carapace from above.

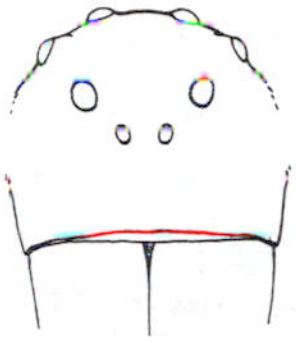

Figure 506. *Peucetia viridans,* face.

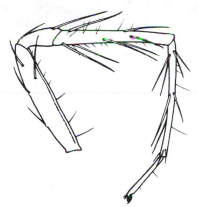

Figure 504. *Oxyopes,* leg.

1a Posterior row of eyes slightly procurved (fig. 505); posterior lateral eyes farther from the anterior laterals than from the posterior medians. The row of anterior lateral eyes wider than the row of posterior medians (fig. 506). Retromargin of cheliceral fang furrow without teeth. (Body bright green in life) . . . .
. . . . *Peucetia* **(2 species)**

Figure 507. *Peucetia viridans,* female.

Figure 507. *Peucetia viridans* (Hentz)

This species is also known under the name *abboti* (Walckenaer). The cephalothorax is highest in the eye region, where it is quite narrow, but broadens out considerably behind. The body is green, bright in life, and usually with red spots in the eye region and over the body as a whole varying in number and size. The legs are paler green to yellow, quite long, provided with very long black spines, and covered with numerous black spots, particularly noticeable on the femora. Length of female 14 to 16 mm; of male 12 to 13 mm.

Found in tall grass, and in California commonly in the flowering heads of wild buckwheat.

Virginia south to Florida and west to California.

1b **Posterior row of eyes strongly procurved (fig. 508); posterior lateral eyes about as far from the anterior laterals as from the posterior medians. The row of anterior lateral eyes subequal in length to the row of posterior medians (fig. 508) or else narrower (fig. 509). Retromargin of cheliceral fang furrow with one tooth. (Body in life not green)** 2

Figure 508. *Oxyopes* face.

Figure 509. *Hamataliwa grisea,* face.

2a **Leg IV longer than III. Eyes of posterior row subequidistant** ....
.... *Oxyopes* (12 species)

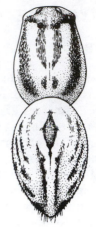

Figure 510. *Oxyopes salticus.*

Figure 511. *Oxyopes salticus,* face and chelicerae.

Figures 510 and 511. *Oxyopes salticus* Hentz

The carapace is mostly yellow with four indistinct longitudinal gray bands. In the female the abdomen is lighter than the carapace, darker along the sides and with a black band on the venter. In the male the abdomen is entirely gray or black, above and below, and with scales that make it appear iridescent. The femora I, II, and III are yellow, with a single narrow jet black line on the ventral side. There is a similar black line extending from each anterior median eye down the clypeus and the front of each chelicera (fig. 511). Length of female 5.7 to 6.7 mm; of male 4 to 4.5 mm.

Throughout the United States but more

common in the East, and apparently uncommon in the Rocky Mountain and Great Basin areas.

Oxyopes aglossus Chamberlin

The face looks much like that of *salticus* with the black lines down the front. There is a distinct black line under femora I and II, but on III and IV there is a series of black dashes broken and indistinct. The carapace is yellow with brown bands near the margins. The abdomen is yellowish white with the sides brown. The venter has a central black band. As in *salticus* the abdomen of the male has iridescent scales. Length of female 4.5 to 6.7 mm; of male 3.9 to 4.8 mm.

Virginia south to Georgia and west to Texas.

Figure 512. *Oxyopes scalaris*, face and chelicerae.

Figure 512. *Oxyopes scalaris* Hentz

The sides of the carapace are reddish-brown, the center yellowish. The abdomen above is darker than the carapace, somewhat lighter on the sides, and with a broad reddish-brown band on the venter from epigynum to spinnerets. The femora I, II, and III each have two reddish-brown broad bands, not distinct lines on the ventral side. There is a similar broad band extending from each anterior lateral eye down the clypeus, and continuing down the front of each chelicera. Length of female 7 to 8 mm; of male 5 mm.

Throughout the United States but more common in the west than is *salticus,* though not as common in the Great Plains area.

Oxyopes apollo Brady

The face looks much like that of *scalaris,* with a brown band down the front of the clypeus and chelicerae. The legs are without the ventral bands on the femora. The carapace is yellowish orange, with a brownish band each side. The abdomen has a central light band, but is brown on the sides. Length of female 4.2 to 6.7 mm; of male 3.4 to 4.4 mm.

Tennessee and Missouri southwest to Arkansas and Texas.

2b **Legs III and IV subequal in length, or III longer than IV. Eyes of the posterior row, in most, with the medians closer to the laterals than to each other (fig. 513). *Hamataliwa* (3 species)**

Figure 513. *Hamataliwa grisea,* face.

Figure 513. *Hamataliwa grisea* Keyserling

The carapace is orange to reddish brown, with, in the female, a whitish band extending back along each side. These bands are lacking in the male. The abdominal dorsum is brown, gray, or black. Length of female 8.7 to 10.9 mm; of male 8.4 to 11.1 mm. Specimens have been collected from against the bark of woody shrubs and trees where they appear to blend with the background.

Florida west across the southern tier of States to California.

FAMILY GNAPHOSIDAE

In these spiders the cephalic part is not sharply set off from the thoracic. The margins of the cheliceral fang furrows are oblique and usually armed. The teeth may be considerably reduced in size to minute denticles, or a single denticle, or the retromargin may have a keeled lamina in place of teeth. The labium is longer than broad, with the endites converging more or less. Usually the endites have an oblique or transverse depression. The legs are generally spinose and the tarsi furnished with scopulae, two claws, and claw tufts. The abdomen is oval and usually rather flattened and in most males provided with a scutum at the anterior end. Many species are uniformly colored, more often dark than light and without markings. Some species have a striking or colorful pattern of lines or spots. The anterior spinnerets are cylindrical, longer and more heavily sclerotized than the posterior, and widely separated. There are 17 genera in our region.

These spiders spin a tubular retreat under stones or in rolled leaves, from which they emerge to hunt.

1a Retromargin of cheliceral fang furrow with a keeled lamina (b) (figs. 514, 515). 2

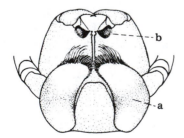

Figure 514. *Gnaphosa* mouthparts from below showing wide serrated lamina (b) on chelicera.

Figure 515. *Callilepis,* left chelicera from below showing narrower, unserrated lamina.

1b Retromargin of cheliceral fang furrow smooth, or with one or more distinct teeth (a) or denticles (fig. 516) 3

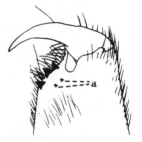

Figure 516. *Drassodes,* left chelicera from below showing distinct teeth on retromargin of fang furrow.

2a Posterior eye row much longer than the anterior, with the medians nearer to each other than to the laterals. Retromargin of chelicera with a wide serrated lamina (see fig. 514 (b)) . . . . *Gnaphosa* (20 species)

GENUS GNAPHOSA

Most members of this genus have the carapace dark brown with black markings along the radial furrows. The abdomen is dark gray to black covered with fine hairs. The scutum on the abdomen of the male is small and indistinct. Most specimens are collected from under stones and old logs on the ground, and in pastures.

Figure 517. *Gnaphosa muscorum.*

Gnaphosa muscorum (L. Koch)
Length of female 8.5 to 15 mm; of male 7.3 to 12 mm.

New England south to Virginia and west through the northern tier of States to the Pacific Coast States. Also, all of Canada to Alaska.

Gnaphosa brumalis Thorell
Similar in general appearance to *muscorum*. Length of female 7.9 to 10 mm; of male 6.8 to 9 mm.

Northern New England and adjacent Canada; the Rocky Mountain States west to Oregon and California, and western Canada to Alaska.

Gnaphosa parvula Banks
Similar in general appearance to *muscorum*. Length of female 7 to 11.5 mm; of male 6 to 6.8 mm.

New England south to West Virginia and west through the northern States to Colorado, then Washington and Oregon. Also, all of Canada to Alaska.

Gnaphosa californicus Banks
Similar in general appearance to the three preceding species but averaging somewhat smaller in size. Length of female 5.4 to 7 mm; of male 5.3 to 6.8 mm.

Rocky Mountain and Pacific Coast States and British Columbia.

Gnaphosa sericata (L. Koch)
This species is smaller and lighter than the preceding ones. The cephalothorax and legs are yellowish brown to orange and the abdomen gray to black. Length of female 4.4 to 7 mm; of male 4 to 6 mm.

Long Island, New York, south to Florida and west to Arizona and Utah.

2b **Posterior eye row but little longer than the anterior, with the eyes equidistant or the medians farther from each other than from the laterals. Retromargin of chelicera with a narrow lamina, without serrated edge (see fig. 515 (b)) . . . . *Callilepis* (7 species)**

Figure 518. *Callilepis pluto,* male.

Figure 518. *Callilepis pluto* Banks

For many years this species had been mistakenly identified as *imbecilla*. The cephalothorax is orange, the legs similar or dark brown. The abdomen is bluish black above with sometimes lighter hairs at the posterior end. Length of female 4.8 to 6.2 mm; of male 3.8 to 5 mm.

This species may be found under stones in pastures.

New England southwest to Alabama, west across the northern tier of States to the Rockies and Pacific Northwest. Also southern Canada.

Callilepis imbecilla (Keyserling)
Similar to *pluto* but the males average much smaller. Length of female 3.7 to 5.6 mm; of male 3 to 3.6 mm.

Ohio to Illinois southwest to Texas, and along the Gulf of Mexico to Florida.

Callilepis eremella Chamberlin
Similar to *imbecilla* but the females average somewhat larger. Length of female 4.5 to 6.1 mm; of male 3.1 to 3.6 mm.

Rocky Mountain States west to the Pacific Coast States.

3a   Tibia IV with one or two median dorsal spines; tibia III with one. Trochanters notched (as in figs. 519 or 523). 4

3b   Tibia IV with no median dorsal spines; tibia III with one or none. Trochanters not notched 6

4a   Tibia IV with two median dorsal spines. Trochanters deeply notched (fig. 519) . . . . *Drassodes* (7 species)

Figure 519. *Drassodes,* deeply notched trochanter.

GENUS DRASSODES

For the most part these spiders are pale yellow to reddish brown. The trochanters are deeply notched. The males lack the dorsal abdominal scutum present in the other genera. Specimens have been collected from under stones in pastures and wooded areas.

Figures 520 and 521. *Drassodes neglectus* (Keyserling)

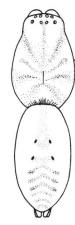

Figure 520. *Drassodes neglectus.*

Figure 521. *Drassodes neglectus,* female, palpal femur and patella.

The palpal femur is provided with three dorsodistal spines as illustrated here. There are two (or more) denticles on the retromargin of the cheliceral fang furrow. The general color is yellow to light gray with a faintly indicated longitudinal stripe on the anterior part of the abdomen and indistinct chevrons on the posterior part. The posterior median eyes are oval, oblique, and much closer to each other than to the laterals. Length of female 8.8 to 13 mm; of male 7.5 to 12 mm.

Found under stones in pastures and wooded areas.

New England south to Virginia, west through the northern tier of States to the Rockies, and to the Pacific Coast States. Also, all of Canada to Alaska.

Figure 522. *Drassodes saccatus*, female, palpal femur and patella.

Figure 522. *Drassodes saccatus* (Emerton)

Similar in general appearance to *neglectus*, but the males average somewhat smaller. Like that species there are two denticles on the retromargin of the cheliceral fang furrow, but the palpal femur is provided (usually) with six dorsodistal spines as illustrated. Length of female 8.2 to 10 mm; of male 7.5 to 10 mm.

New England southwest to Kansas and the Rocky Mountain States, then west to the Pacific Coast.

Drassodes gosiutus Chamberlin
Although there are three dorsodistal spines on the palpal femur as with *neglectus*, there is only

one denticle on the retromargin of the cheliceral fang furrow. Length of female 7.8 to 11 mm; of male 7.7 to 9.7 mm.

Southern New England southwest to Texas and west to the Rocky Mountain States and Alberta.

4b **Tibia IV with one median dorsal spine. Trochanters only slightly notched (fig. 523)** 5

Figure 523. *Herpyllus*, slightly notched trochanter.

5a **Height of clypeus twice or more the diameter of an anterior median eye. Posterior median eyes separated from each other by one and a half to two times a diameter and only one diameter from the posterior laterals. Two black longitudinal bands on carapace and two or three black bands running the entire length of the abdomen (fig. 524 and 525)** . . . . *Cesonia* (3 species)

Figure 524. *Cesonia bilineata*.

Figure 524. *Cesonia* bilineata (Hentz)

Running the entire length of both carapace and abdomen are two broad black bands which alternate with three white ones. The cephalothorax is much narrowed in front and the two eye rows are virtually straight. Length of female 6 to 8 mm; of male 5 to 6 mm.

Found under old leaves and stones in wooded areas. Runs very rapidly when disturbed.

New England south to Georgia and west to Nebraska.

Figure 525. *Cesonia classica.*

Figure 525. *Cesonia classica* Chamberlin

Similar to *bilineata,* but with three black bands running the length of the abdominal dorsum. Length of female 5 to 6 mm; of male 3.9 to 4.6 mm.

California and Arizona.

5b Height of clypeus not more than the diameter of an anterior median eye. Posterior eyes subequidistant. Carapace without a white band, and abdomen not marked as above . . . .
. . . . *Herpyllus* (25 species)

GENUS HERPYLLUS

In this genus the abdomen is mousey gray in color, sometimes with a light band as illustrated for *ecclesiasticus*. Males have a conspicuous brown abdominal scutum.

Figure 526. *Herpyllus ecclesiasticus.*

Figure 526. *Herpyllus ecclesiasticus* Hentz. Parson Spider

The carapace and legs are chestnut brown with indistinct gray markings. The abdomen is gray to black above with a light median band on the basal two-thirds, and a light spot behind. The body is covered with silky gray hairs. Length of female 8 to 13 mm; of male 5.5 to 6.5 mm.

Found under stones and old boards on the ground in wooded areas, and hibernates under loose bark.

New England and adjacent Canada south to Georgia and west to Oklahoma and Colorado.

Herpyllus propinquus (Keyserling)
The carapace and legs are brownish orange with some dusky blotches. The abdomen is gray with the characteristic light band as in *ecclesiasticus*. In the male the light area is not as clear. Length of female 9.3 to 11 mm; of male 6 to 7 mm.

Montana to Arizona and west to Oregon and California.

Herpyllus blackwalli (Thorell)
This species is also known under the name *Scotophaeus blackwalli*. The carapace and legs are orange brown, covered with gray hairs. The abdomen is grayish brown covered with a fine pubescence. Length of female 9 to 12 mm; of male 7 to 10 mm. A European species probably imported and often found in houses.

Gulf Coast and Pacific Coast States.

6a Distal end of metatarsi III and IV provided with a ventral comb (fig. 527)..... 7

Figure 527. *Zelotes,* preening comb on metatarsus IV.

6b Without such a comb................8

7a Posterior median eyes larger than the laterals, oval in most; the posterior row slightly procurved and the medians closer to each other than to the laterals ....*Drassyllus* (46 species)

GENUS DRASSYLLUS

Most of the species in this large genus occur in the western states. The colors vary from orange to dark brown on the cephalothorax, and gray to black on the abdomen. The darker ones resemble *Zelotes* with which they have been confused. The spiders are found under leaves, stones, and logs on the ground.

Figure 528. *Drassyllus depressus.*

Figure 528. *Drassyllus depressus* (Emerton)

The cephalothorax is orange to dark brown and the abdomen gray to black without markings. Length of female 4.7 to 6 mm; of male 5 mm.

New England and adjacent Canada south to Georgia and west to Arizona and Colorado; also Oregon.

Drassyllus niger (Banks)
This is one of the darkest species with the carapace and legs dark brown and the abdomen almost black. Length of female 7 to 9 mm; of male 5 to 6 mm.

New England west through Pennsylvania to Minnesota.

Drassyllus insularis (Banks)
Similar to the preceding species but smaller. Length of female 5 to 6.5 mm; of male 4.8 to 5 mm.

Utah and Arizona west to California.

Drassyllus fallens Chamberlin
Length of female 5 to 6.5 mm; of male 4.3 to 4.6 mm.

New England south to North Carolina and west to Wisconsin.

Drassyllus virginianus Chamberlin
Length of female 7.4 to 8 mm; of male 7 mm.

New England south to Virginia and west to Wisconsin.

Drassyllus frigidus (Banks)
Length of female 4.8 to 5.2 mm; of male 4.3 to 4.7 mm.

New England south to Georgia, west through Pennsylvania to Ohio.

Drassyllus rufulus (Banks)
Length of female 5.7 to 8.5 mm; of male 6 mm.

New England and adjacent Canada south to Maryland and west to Colorado.

Drassyllus aprilinus (Banks)
Length of female 4.3 to 5.7 mm; of male 5.4 to 6 mm.

New England south to Virginia and west to Kansas.

Drassyllus agilis (Bryant)
This species is much lighter than the usual, with the carapace a golden brown, and the abdomen gray. Length of female 6.7 mm; of male 5 mm.

Texas west to California.

7b Posterior median eyes hardly, if at all, larger than the laterals, and circular in most (rarely oval). Posterior eye row straight (rarely procurved) and eyes equidistant. . . . .
. . . . *Zelotes* (30 species)

GENUS ZELOTES

In general the colors vary from brownish gray to black without markings, (fig. 529) and formerly many species were confused with *subterraneus*. They closely resemble also the darker species of *Drassyllus* with which many have been confused.

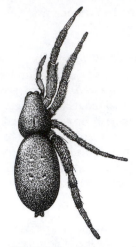

Figure 529. *Zelotes subterraneus,* female.

Figure 530. *Zelotes hentzi.*

Figure 530. *Zelotes hentzi* Barrows

Length of female 5.8 to 8 mm; of male 4.6 to 6.3 mm.

Eastern, southern and central states and southern Canada.

Zelotes subterraneus (C.L. Koch)
Length of female 6 to 9 mm; of male 5 to 6.5 mm.

Throughout the northern states and adjacent Canada southwest to Kansas.

Zelotes puritanus Chamberlin
Length of female 6.7 to 7 mm; of male 4.4 to 5 mm.
New England across the northern states to Washington.

Zelotes anthereus Chamberlin
Length of female 5.6 to 8.7 mm; of male 4.8 to 6 mm.
California.

Zelotes rusticus (L. Koch)
This species is more lightly colored than is usual for members of this genus. The carapace and legs are orange. The abdominal scutum in the male is narrow and short. The dorsum is creamy with a pale gray pubescence. Length of female 9 to 9.8 mm; of male 5.6 to 7.1 mm.
Throughout the United States.

8a   **Carapace in front narrowed to half its greatest width. Median ocular area slightly wider behind than in front. Abdomen with white areas. . . . .**
. . . .*Poecilochroa* (25 species)

GENUS POECILOCHROA

Of the species given here the first, *montana,* is devoid of a median dorsal spine on tibia III. The other four species possess such a spine and had formerly been placed in the genus *Sergiolus.*

Figure 531. *Poecilochroa montana.*

Figure 531. *Poecilochroa montana* Emerton

The carapace is chestnut to mahogany brown, with smoky markings and sparse white pubescence. The femora are similar but the distal leg segments are orange. The abdomen is black except for a pair of white spots near the middle of its length. This species differs from the other four included here in that it lacks a median dorsal spine on tibia III. Length of female 7.3 to 8.6 mm; of male 5.8 mm.
New England across the northern states to Rocky Mountain States, Alberta and the Pacific Coast States.

Figure 532. *Poecilochroa capulata,* female.

Fig 532. *Poecilochroa capulata* (Walckenaer)

This species is also known under the name *variegatus* (Hentz). The carapace is bright orange, a little darker toward the eyes, but black on the posterior declivity (which is overhung by the abdomen). The legs have femora I and II black, but are elsewhere orange except for black rings at the distal end of femur IV and both ends of tibia IV. The abdomen is black with three transverse bands of white scales and a T-shaped mark extending forward from the center of the middle band. The scutum of the male is just a little over half the length of the dorsum. Length of female about 10 mm; of male 5.5 to 7 mm.

Found under stones and dead leaves in woods and sandy places.

New England and adjacent Canada south to Georgia and west to Oklahoma and Nebraska.

Figure 533. *Poecilochroa famula,* male.

Figure 533. *Poecilochroa famula* (Chamberlin)

The carapace is orange brown, not as bright as in *capulata*. The abdomen is black with three transverse bands of white scales. The first and second are connected along the sides, and the middle band is narrower than the other two. The scutum in the male extends back not quite two-thirds the length of the dorsum. Length of female 6.7 to 8.5 mm; of male 5 to 5.7 mm.

New England and New York south to the District of Columbia.

Figure 534. *Poecilochroa decorata,* male.

Figure 534. *Poecilochroa decorata* (Kaston)

In general appearance and markings this species resembles *capulata,* but neither the posterior declivity of the carapace nor the proximal end of tibia IV are black. In addition, in the male the central T-shaped spot is not so distinct and the scutum is larger than in that species. Length of female 5.4 to 6.7 mm; of male 4.7 mm.

New England west to South Dakota.

Poecilochroa ocellata (Walckenaer)
This species is quite similar in appearance to *decorata* and *capulata* from which it may be distinguished by the lack of black markings on the legs. Length of female 6.2 mm; of male 4.2 to 5 mm.

Pennsylvania south to Florida and west through Texas to California.

8b Carapace in front not narrower than two-thirds to three-fifths its greatest width. Median ocular area as wide in front as behind. Abdomen unspotted. . . 9

9a Posterior median eyes oval and oblique, closer to each other than to the laterals . . . . *Haplodrassus* (9 species)

Figure 535. *Haplodrassus signifer.*

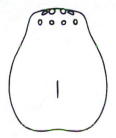

Figure 536. *Orodrassus coloradensis.*

Figure 535. *Haplodrassus signifer* (C.L. Koch)

The carapace and legs are yellowish orange, the abdomen is gray. Length of female 7.5 to 11 mm; of male 7 to 9 mm.

Found under stones and boards in pastures.

Throughout the United States and Canada.

Haplodrassus hiemalis (Emerton)
In general appearance similar to *signifer.* Length of female 7 to 8.5 mm; of male 5.9 to 8 mm.

New England and adjacent Canada west to Wyoming and Colorado.

Haplodrassus bicornis (C.L. Koch)
Similar to the two preceding species but smaller. Length of female 4 to 6.5 mm; of male 3.6 to 4.6 mm.

New England south to Virginia and west to the Rockies. In Canada from Ontario to British Columbia.

Haplodrassus chamberlini Platnick & Shadab
Length of female 5.8 to 8.6 mm; of male 3.9 to 5.6 mm.

Nebraska south to Oklahoma west to the Rockies and southern California.

9b   Posterior median eyes circular, and the eyes of the posterior row subequidistant . . . . *Orodrassus* (4 species)

Figure 536. *Orodrassus coloradensis* (Emerton)

The cephalothorax and legs are orange to brown, the abdomen gray without markings, but covered with dense pubescence. Length of female 9 to 10 mm;; of male 7.3 to 8.5 mm.

Rocky Mountain region from New Mexico to Manitoba west to British Columbia and Oregon.

Orodrassus assimilis (Banks)
Similar in appearance to *coloradensis,* but averaging somewhat larger in size. Length of female 8.4 to 10.6 mm; of male 8 to 9.2 mm.

The southern Rocky Mountain region west to California and Oregon.

FAMILY HOMALONYCHIDAE

This is a small family restricted to our southwest, and with a single genus, *Homalonychus,* containing three species.

Figure 537. *Homalonychus theologus.*

Figure 537. *Homalonychus theologus* Chamberlin

The general color is orange brown with a mottling of dark brown spots on body and legs. Most specimens have the body covered with sand grains, which makes the spider difficult to see against its background. This is especially the case as the spider assumes a position lying with its entire body and legs flat against the substratum. The coxae IV are widely separated by a sternum which is as wide as long. All tarsi are scopulate and the tarsal claws are untoothed. Length of female 8.5 to 10 mm; of male 6.5 to 7 mm.

Arizona and California, under stones and on sand.

FAMILY CLUBIONIDAE

This is a large family of two-clawed hunting spiders commonly encountered on foliage or on the ground, where they may make tubular retreats in rolled up leaves, or under stones, in litter, etc. There is little difference between the sexes though the males are slightly smaller, often with the chelicerae longer and narrower and the legs somewhat longer. There are 24 genera in our region.

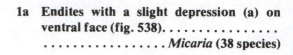

1a Endites with a slight depression (a) on ventral face (fig. 538). . . . . *Micaria* (38 species)

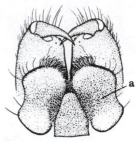

Figure 538. *Micaria,* mouthparts from below showing depression across endites.

GENUS MICARIA

Because of the depression on the endites, and the fact that the posterior median eyes often appear oval, as in many gnaphosids, some workers in recent years have placed this genus in the Gnaphosidae. The median thoracic groove is lacking or only faintly indicated and the body is covered with flattened scales, usually brightly colored and iridescent, but unfortunately often lost in preserved material. In many cases the abdomen shows a constriction or depression near its anterior end but the degree of depression varies even within the same species, especially in females where the state of gravidity is presumably a factor (fig. 539). These spiders run very rapidly over the ground in dry areas, and resemble ants. From the similar appearing *Castianeira* species they may be distinguished by characters discussed under that genus.

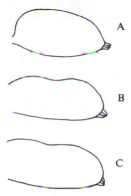

Figure 541. *Micaria longipes,* abdomen of female.

Figure 539. *Micaria,* abdomens of three different females, from the side, to show variation in the amount of dorsal depression.

Figure 541. *Micaria longipes* Emerton

This species is quite similar to the preceding. The anterior half of the abdomen has two pairs of white spots thinner than in *aurata,* and the posterior half is black, usually without any trace of a chevron pattern. Length of female 5 to 6 mm; of male 5 mm.

Northern states west to Utah.

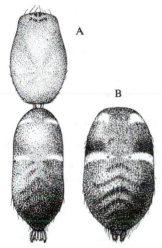

Figure 540. A, *Micaria aurata* male; B, *Micaria aurata,* abdomen of female.

Micaria gertschi Barrows & Ivie
The carapace is dark brown, sparsely covered with white scales. The abdomen is black above, evenly covered with iridescent scales, and with the dorsum showing two small white spots, and with a larger median pair of bars continuous down the sides. Length of female 3.5 mm; of male 3.1 mm.

Michigan and adjacent Canada west to Colorado.

Figure 540. *Micaria aurata* (Hentz)

The general color is light yellow-brown with gray hairs and also iridescent scales. This species is quite similar to the next. The four white spots on the anterior half of the abdomen are thicker than in *longipes* and the posterior half of the abdomen shows a chevron pattern. Length of female 5.2 to 6.7 mm; of male 5 to 6 mm.

New England to Texas and Nebraska, but more common in the South.

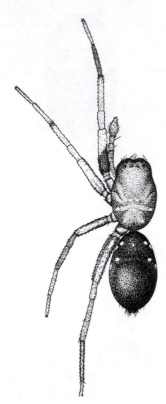

Figure 542. *Micaria laticeps*.

Figure 542. *Micaria laticeps* Emerton

The general color is dark brown to black with iridescent scales all over the body, and spots of white scales on carapace and abdomen as shown in the drawing. In both sexes there is a slight depression just in front of the middle of the abdomen. The basal half of femur I is dark brown while the distal half is white like the patella and tibia. Length of female 2.8 mm, of male 2.7 mm.

This spider occurs under stones in pastures.

New England.

1b   Endites without a depression 2

2a   Leg I longer than IV 3

2b   Leg IV longer than I 5

3a   Cephalothorax orange-brown to reddish above and below, smooth and shining without hairs. Anterior legs darkest, other legs become increasingly paler towards the rear. Sternum distinctly margined (a) (fig. 543) . . . .
. . . . *Trachelas* (6 species)

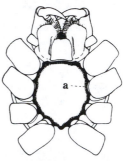

Figure 543. *Trachelas*, cephalothorax from below.

GENUS TRACHELAS

All our species resemble one another closely. The sternum and carapace are reddish brown and thickly covered with tiny punctures. The abdomen is pale yellow to light gray, slightly darker in the anterior median area. The legs are darker from first to fourth. Specimens have been collected from under the loose bark of trees, and from rolled up leaves, and on many occasions have been found in autumn walking about inside houses.

Trachelas tranquillus (Hentz)
This species has also been known under the name *ruber* Keyserling. The anterior eyes are separated by a diameter of one and the height of the clypeus is equal to the same diameter. Length of female 6.8 to 10 mm; of male 5 to 6.1 mm.

New England and adjacent Canada southwest to northern Alabama and Georgia, and west to Minnesota and Kansas.

Figure 544. *Trachelas*.

Trachelas pacificus Chamberlin and Ivie

Similar to *tranquillus* with respect to the eyes and clypeus. However, the females average slightly smaller and the males slightly larger in size. Length of female 6 to 7.7 mm; of male 5.3 to 6.5 mm.

California and Nevada.

Trachelas similis F.O.P.-Cambridge

The anterior eyes are closer together than in the preceding two species, being separated by less than a diameter of one. Likewise the clypeus height is less than the diameter of one. Length of female 5.8 to 7.2 mm; of male 4.8 to 6 mm.

Virginia south to Florida and west to Texas.

Trachelas deceptus (Banks)

The eyes in the anterior row are separated by at least a diameter of one, and the height of the clypeus is equal to, or less than, the diameter. This species is much smaller than the others and differs also in having the posterior row of eyes straight rather than recurved. Length of female 3.4 to 4.2 mm; of male 3.1 to 4.1 mm.

New York south to Florida and west to Michigan and Texas; also in Colorado and California.

3b **Cephalothorax light yellow to greenish, brownish at most only in eye region. All legs about the same color. Sternum not distinctly margined** **4**

4a **Anterior median eyes much the largest. Posterior medians slightly nearer to posterior laterals than to each other. Thoracic groove present (fig. 545)** . . . . *Strotarchus* (2 species)

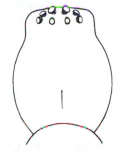

Figure 545. *Strotarchus*, carapace.

Figure 545. *Strotarchus piscatoria* (Hentz)

The carapace is yellowish brown, darker in the eye region. The abdomen is brown to gray above. Length of female 8 to 9 mm; of male 7 to 8 mm.

This species spins a silken retreat under stones.

New England south to Florida and west to Alabama.

4b **Eyes subequal. Posterior medians slightly nearer to each other than to the posterior laterals. Thoracic groove lacking (fig. 546)** . . . . *Chiracanthium* (2 species)

Figure 546. *Chiracanthium*, carapace.

GENUS CHIRACANTHIUM

The members of this genus are ordinarily found walking about on foliage, where they spin silken retreats. On occasion they have been the cause of envenomation to humans, particularly gardeners who accidentally disturb them. The body is pale yellow to pale green with abdomen changing color somewhat in accordance with the color of the food imbibed. The chelicerae are dark brown.

Chiracanthium inclusum (Hentz)

Length of female 4.9 to 9.7 mm; of male 4 to 7.7 mm.

Throughout the United States except for the most northern tier of States.

Chiracanthium mildei L. Koch

This species has been found in houses more frequently than *inclusum*. Length of female 7 to 10 mm; of male 5.8 to 8.5 mm.

Presumably introduced from Europe, and in recent years found in New England, New York, New Jersey, Alabama, Missouri, and Utah.

5a Labium wider than long, or at most only half as long as endites, which are not constricted in the middle, but either as wide as at end or wider. 7

5b Labium longer than wide and exceeding half the length of endites, which are narrower in the middle than at the end 6

6a Femur I with one distal prolateral spine. Dorsum of abdomen pale yellowish brown to reddish brown . . . .
. . . . *Clubiona* (59 species)

GENUS CLUBIONA

In these spiders the colors are generally white, cream, or tawny, with darker brown at the cephalic end and on the chelicerae. The body is covered with short hairs which give it a silky reflection.

Most are unmarked (fig. 549) and hence difficult to distinguish except by genitalia or secondary sex characters. The chelicerae are stout in females, in males generally more slender, longer and tapering and sometimes with sharp ridges, or keels along the anteromedial face, the lateral face, or both. The tarsal claws (a) are long and the claw tufts (b) very conspicuous (fig. 547).

These spiders run rapidly over plants and make tubular silken retreats. They spend the winter under bark and stones.

Figure 547. *Clubiona,* claws and claw tufts.

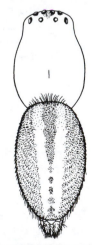

Figure 548. *Clubiona riparia.*

Figure 548. *Clubiona riparia* L. Koch

In the male the chelicerae are keeled on their anteromedial face. The abdomen is reddish brown with a pair of yellow lines enclosing a central dark strip broken into spots behind. Length of female 5.4 to 8.7 mm; of male 4.4 to 7.4 mm.

The spider lives in tall grass, a blade of which is folded by the female to make a three-sided chamber. This is lined with silk and is used as a place of concealment for the egg sac.

New England west to Illinois and south to Maryland. In Canada northwest to Alaska.

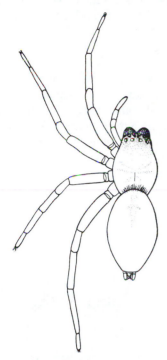

Figure 549. *Clubiona obesa.*

Figure 549. *Clubiona obesa* Hentz

In the male the chelicerae are keeled on both the anteromedial and lateral faces. Like most other species in the genus this one is without markings. Length of female 6.7 to 11.4 mm; of male 6.7 to 8 mm.

Most common of the larger species.

New England south to North Carolina and west to Nebraska. Also, adjacent Canada west to Manitoba.

Clubiona moesta Banks

In the male the chelicera has a pronounced keel on the lateral face, and a less conspicuous ridge on the anteromedial face. Length of female 4.7 to 6.4 mm; of male 4 to 6 mm.

New England west to Colorado. In Canada from Manitoba to Alaska.

Clubiona maritima L. Koch

This species has long been known under the name *tibialis* Emerton. The male has very long, slender chelicerae provided with a very conspicuous keel on the anteromedial face and a much weaker one on the lateral face. Length of female 6.4 to 9.8 mm; of male 5.7 to 8 mm.

New England and adjacent Canada to Georgia and west to Nebraska.

Clubiona kastoni Gertsch

The chelicerae of the male are not keeled. Length of female 4.3 to 5.4 mm; of male 3.3 to 4.4 mm.

New England and New York south to Tennessee west to Kansas; also California and Oregon as well as southern Canada.

Clubiona abbotii L. Koch

As in *kastoni* the chelicerae of the male are not keeled. This is the most common of the smaller species. Length of female 4 to 5.4 mm; of male 3.7 to 4.4 mm.

New England south to Georgia and west to the Mississippi River and in Canada west to Alberta.

6b Femur I with two distal prolateral spines. Dorsum gray to yellow gray, and usually with gray markings . . . .
. . . . *Clubionoides* (5 species)

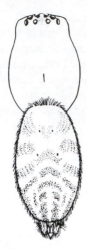

Figure 550. *Clubionoides excepta.*

Figure 550. *Clubionoides excepta* (L. Koch)

There are several rows of black spots on the abdomen, and there is nearly always a pair on the venter just in front of the spinnerets. Length of female 6.7 to 7.4 mm; of male 4.5 to 6.5 mm.

This spider is commonly encountered under dead leaves, stones, and loose bark.

New England and adjacent Canada south to Georgia and west to Kansas and Nebraska.

7a Tibia I with fewer than four pairs of ventral spines. 8

7b Tibia I with four or more pairs of ventral spines 9

8a Carapace rather convex, about two-thirds as wide as long and evenly colored orange, dark brown, or black. Median eyes of both rows slightly larger than, and farther apart from each other than from, the laterals. Abdomen mostly dark with markings in the form of distinct lines, or spots . . . .
. . . . *Castianeira* (25 species)

GENUS CASTIANEIRA

These spiders run about over the ground and may resemble large ants or mutillid wasps. From species of *Micaria,* which also resemble ants, they can be separated by the following characters: the thoracic groove is well marked; the tibiae I and II have two or three pairs of ventral spines; both margins of the fang furrows of the chelicera have two teeth; the labium is wider than long and the endites lack an oblique depression (fig. 551). These spiders resemble in form and manner of movement large carpenter ants, and they have been found associated with the ants. While ordinarily they move about slowly, like the ants, raising and lowering their abdomens and their front legs (which simulate the antennae of ants), they may run very rapidly when disturbed. They are found under stones in pastures and under logs in wooded areas.

The egg cocoons are usually attached to the underside of a stone, are flattened discs, the free surface often of a beautiful opaline or pearly luster, and are tough and difficult to tear open.

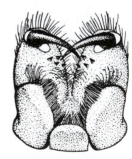

Figure 551. *Castianeira,* mouthparts from below.

Figure 552. *Castianeira cingulata.*

Figure 552. *Castianeira cingulata* (C.L. Koch)

The carapace and abdomen are a dark mahogany brown to black. The abdomen has two transverse white bands and there are black and white stripes on the femora. Length of female 6.7 to 8 mm; of male 6.4 to 7 mm.

New England and adjacent Canada south to Florida and northwest through Arkansas to South Dakota.

Figure 553. *Castianeira longipalpus.*

Figure 553. *Castianeira longipalpus* (Hentz)

Similar in appearance to *cingulata,* but with additional transverse white bands on the abdomen. These bands are often indistinct, though usually at least one is visible near the posterior end. Length of female 7 to 10 mm; of male 5.5 to 6.1 mm.

New England and adjacent Canada south to Florida, west to Oklahoma and North Dakota; and in the Pacific Northwest.

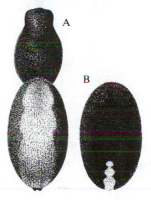

Figure 554. *Castianeira descripta,* showing two varieties of abdominal markings.

Figure 554. *Castianeira descripta* (Hentz)

The carapace and abdomen are deep mahogany brown to black. On the abdomen are red spots often restricted to the posterior end but sometimes extending forward to the anterior end. The legs have the femora dark like the carapace, but the distal segments are lighter, especially on legs I and II. Length of female 8 to 10 mm; of male 6.2 to 7.6 mm.

New England and adjacent Canada to Florida west to Texas, Oklahoma and Iowa.

Figure 555. *Castianeira occidens.*

Figure 555. *Castianeira occidens* Reiskind

This western species has long been confused with *descripta,* but besides differing in genitalia characters shows a light longitudinal band along the middle of the carapace. Length of female 7.5 to 9 mm; of male 6.4 to 6.9 mm.

Utah and New Mexico west to California.

Figure 556. *Castianeira gertschi,* male.

Figure 556. *Castianeira gertschi* Kaston

The carapace is bright orange with a thin black, marginal line. The legs are about the same color but with femur I slightly darker and the distal segments yellow. The distal segments of leg IV are darker than the carapace. The abdomen has two transverse white stripes (or two pairs of white spots) and is orange on the anterior half but becomes increasingly darker toward the posterior end, where it is chestnut brown. Length of female 5.1 to 6.3 mm; of male 4.5 to 5.5 mm.

New England and adjacent Canada south to Florida, west to Texas and Oklahoma.

Figure 557. *Castianeira amoena.*

Figure 557. *Castianeira amoena* (C.L. Koch)

The carapace is orange, and the abdomen orange with transverse white bands as shown in the illustration. There are white hairs scattered on the anterior half of the dorsum, and laterally between the black bands. The legs are annulate. Length of female 7 to 8.8 mm; of male 5.7 to 6.8 mm.

North Carolina south to Florida, west to Texas and Kansas.

8b Carapace rather flat, about three-fourths as wide as long, usually orange brown with dusky blotches, and with a black marginal stripe. Abdomen with a pattern of light and dark spots and blotches. . . . . *Agroeca* (6 species)

GENUS AGROECA

In color pattern the species resemble one another fairly closely so that it is difficult to distinguish them on this basis alone. The carapace is bordered by a black marginal stripe becoming less conspicuous on the cephalic part. Extending along the radial furrows from the thoracic groove, not quite to the edges of the carapace, are dusky blotches. The abdomen is orange brown with gray blotches.

These spiders, and the others of the genus, may sometimes be mistaken for gnaphosids because of the fact that the anterior spinnerets are at times somewhat farther apart than is usual for the members of the Family Clubionidae.

Specimens are generally found running about under leaves on the ground and can be sifted from debris.

Figure 559. *Agroeca trivittata.*

Figure 559. *Agroeca trivittata* Keyserling

As in *emertoni* the metatarsi I and II are provided with only two pairs of ventral spines, but the patellae III and IV of the male each have a retrolateral spine. Length of female 5.8 to 7.2 mm; of male 4 to 5 mm.

Utah, Arizona and the Pacific Coast States.

Figure 558. *Agroeca emertoni.*

Figure 558. *Agroeca emertoni* Kaston

The metatarsi I and II are provided with only two pairs of ventral spines which are hardly longer than the thickness of the metatarsus. Patellae III and IV of the male lack a retrolateral spine. Length of female 4.5 mm; of male 4.2 mm.

New England.

Figure 560. *Agroeca ornata.*

Figure 560. *Agroeca ornata* Banks

The markings along the radial furrows (typical for the genus) are very faint in this species. The metatarsi I and II are provided with three pairs of ventral spines, and the patellae III and IV of the male lack a retrolateral spine. Length of female 5 to 6.6 mm; of male 4.4 mm.

Northern states and southern Canada west to the Rockies.

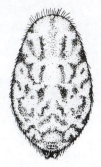

Figure 561. *Agroeca pratensis.*

Figure 561. *Agroeca pratensis.* Emerton

The metatarsi I and II are provided with three pairs of ventral spines, and the patellae III and IV of the male have a retrolateral spine. Length of female about 7 mm; of male about 5.5 mm.

New England and adjacent Canada south to Georgia and west to the Rockies.

9a   Sternum not prolonged behind between coxae IV which are nearly contiguous ... . *Liocranoides* (5 species)[4]

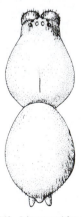

Figure 562. *Liocranoides sober.*

Figure 562. *Liocranoides sober* Chamberlin

The cephalothorax and legs are orange brown, and the abdomen gray. Length of female 13 mm; of male 10 mm.

California.

9b   Sternum prolonged behind between coxae IV, so that these coxae are widely separated 10

10a   Carapace yellow to brown with a black marginal stripe and dark median stripes or spots (fig. 563). Eye rows subequal in width. . . . . *Phrurotimpus* (8 species)

Figure 563. *Phrurotimpus minutus* carapace.

GENUS PHRUROTIMPUS

These spiders usually have iridescent scales on the body. They can run very swiftly and when they stop, they usually flex their legs at the "knee" so as to draw them up over the cephalothorax, hiding the latter. The cocoons are flattened discs, red in color, and tightly fastened to the underside of stones. They are not guarded by the mother.

4. With the possibility of many more species soon to be described.

Figure 564. *Phrurotimpus alarius.*

10b Carapace uniformly dark shiny chestnut or black, without conspicuous contrasting markings. First eye row slightly though distinctly narrower than second . . . .
. . . . *Scotinella* (31 species)[5]

Figure 565. *Scotinella formica* carapace.

Figure 564. *Phrurotimpus alarius* (Hentz)

The general color is yellowish with black markings, with nearly always two gray spots near the middle of the venter, and only slightly, if at all, iridescent. Legs III and IV have black spots, tibia I has five or six pairs of ventral spines. The male lacks a brush of hairs under tibia I. Length of female 2.5 to 2.8 mm; of male 2 to 2.3 mm.

These spiders live among stones and dead leaves on the ground.

New England and adjacent Canada south to Georgia and west to the Rocky Mountain States.

Phrurotimpus borealis (Emerton)

Very similar in appearance to *alarius,* but with the abdomen darker and more iridescent, especially in the male. Also, legs III and IV lack black spots, tibia I has 7 or 8 pairs of ventral spines, and the male has a brush of hairs under tibia I. Length of female 2.4 to 3.6 mm; of male about 2.8 mm.

These spiders live among stones and dead leaves on the ground.

New England west to the Mississippi and south to North Carolina. Also southern Canada.

Figure 565. *Scotinella formica* (Banks)

In both sexes the entire dorsum of the abdomen is covered with a shiny scutum. The carapace is dark brown to black, the abdomen the same or somewhat darker. Length of female 2.1 to 2.8 mm; of male 2.3 to 2.6 mm.

These spiders have been found living with ants.

New England to North Carolina and west to Oklahoma.

Scotinella redempta (Gertsch)

Similar in appearance to *formica.* Length of female 2.5 to 3.6 mm; of male 2.1 mm.

Virginia to Tennessee and Alabama and west to Kansas.

FAMILY ANYPHAENIDAE

These spiders are similar in general appearance to *Clubiona* and related genera in the Clubionidae. Besides having the spiracular furrow far forward they have the claw tufts consisting of lamelliform hairs (fig. 566). They are usually found hunting their prey on foliage. There are five genera in our region.

5. With the possibility of perhaps 10 additional species soon to be described.

Figure 566. *Aysha,* tip of tarsus showing claw tufts composed of lamelliform hairs.

1a **Spiracular furrow (b) much nearer to epigastric furrow (c), only about one-fifth to one-fourth the distance from the latter to the spinnerets (fig. 567). Anterior eyes subequal** . . . .
. . . . *Aysha* **(6 species)**

Figure 567. *Aysha,* venter.

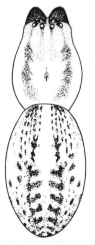

Figure 568. *Aysha gracilis.*

Figure 568. *Aysha gracilis* (Hentz)

The general color is yellow, darker at the anterior end; brown on the chelicerae. There is a pair of grayish bands on ths carapace and reddish brown to black spots forming two indistinct rows on the abdomen. Length of female 6.4 to 7 mm; of male 5.7 to 6.5 mm.

New England south to Florida and west to Texas and Kansas, frequently in houses.

Aysha incursa (Chamberlin)
Similar to *gracilis.* Length of female 5.7 to 7 mm; of male 5.5 to 6.1 mm.

Utah and Arizona west to California.

Aysha velox (Becker)
The carapace is marked like that in *gracilis,* but the abdomen lacks the dark markings. Length of female 8.4 mm; of male 7.3 mm.

North Carolina south to Florida and west to Arkansas and Texas.

1b **Spiracular furrow about midway between epigastric furrow and base of spinnerets (fig. 569). Anterior median eyes smaller than posterior medians** **2**

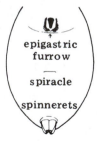

Figure 569. *Anyphaena,* venter.

2a **Tibia I about one and a half times as long as carapace** . . . .
. . . . *Wulfila* **(5 species)**

Figure 570. *Wulfila saltabunda.*

Figure 570. *Wulfila saltabunda* (Hentz)

The general color is white with a pair of broken gray bands on the carapace and several rows of gray spots on the abdomen. The anterior legs are very long, twice or more the length of the entire body. Length of female 3.7 to 4.2 mm; of male 2.9 to 3.5 mm.

New England and adjacent Canada west to Nebraska and Texas.

Wulfila alba (Hentz)
Like *saltabunda* but without the dark markings. Length of female 4 mm; of male 3.6 mm.

Maryland south to Florida and west to Illinois and Texas.

2b Tibia I not, or hardly, longer than carapace **3**

3a Carapace usually with a pattern of two darkly pigmented longitudinal bands, one on either side of the midline (fig. 571). All leg segments approximately concolorous. (Males with chelicerae not elongated) . . . . *Anyphaena* (21 species)

GENUS ANYPHAENA

The species are all more or less similar in color and general appearance. They are pale yellow to white with two faint gray bands running lengthwise on the carapace and several transverse rows of gray spots on the abdominal dorsum. The males of some species have spurs or hirsute knobs, or both, on the underside of coxae II, III, and IV; in other species these are missing.

These spiders are generally found under logs and forest floor litter, and in grass and bushes.

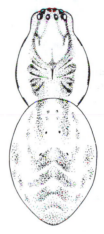

Figure 571. *Anyphaena celer.*

Figure 571. *Anyphaena celer* (Hentz)

The general color is pale yellow to white with two faint gray lines on the carapace and several transverse rows of gray spots on the abdomen. In the male, coxae III and IV are thickly set with numerous short, thick bristles giving the area the appearance of a rasp or stiff brush and femur III is thickened toward its distal end. Length of female 5 to 6 mm; of male 4 to 5 mm.

New England and adjacent Canada south to Alabama and west to Texas and Wisconsin.

Anyphaena maculata (Banks)
Very similar to *celer,* particularly with respect to the modifications on the legs of the male.

Length of female 4.5 to 5.5 mm; of male 3.7 to 4.7 mm.

New York south to Georgia and west to Louisiana and Missouri.

Anyphaena fraterna Banks

The male has a knob on each coxa I, in addition to those on the other coxae. On coxa III the posterior spur is gently curved and sharp, and not bifid. Length of female 5 to 5.7 mm; of male 4.3 to 5 mm.

New England south to Florida and west to Kansas and Texas.

Anyphaena pectorosa L. Koch

The male has coxa II with a low hirsute knob. Coxa III has two processes, of which the anterior is a high hirsute blunt tooth, and the posterior a smooth higher bifid spur. Coxa IV has a gently outcurved smooth spur. Also, there is a low hirsute knob at the center of the sternum. Length of female 5 to 5.5 mm; of male 4.7 to 5.4 mm.

New England south to Florida and west to Missouri and Texas.

Anyphaena pacifica (Banks)

The male differs from those of the three preceding species in that it completely lacks any modifications on the coxae. Length of female 7 mm; of male 5 to 6 mm.

Rocky Mountain States west to the Pacific Coast States and British Columbia.

Anyphaena aperta (Banks)

This is the only species in which the metatarsi I and II are provided with but a single pair of ventral spines; all the other species included here have two pairs. Length of female 5.8 mm; of male 4.3 mm.

British Columbia south to California.

3b Carapace without the darkly pigmented bands. Leg femora darker than the other leg segments. (Males with chelicerae porrect, extending forward considerably) (fig. 572) . . . .
. . . . *Teudis* **(3 species)**

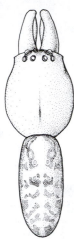

Figure 572. *Teudis mordax*, male.

Figure 572. *Teudis mordax* (O.P.-Cambridge)

The carapace is light reddish brown, smooth and shining. The chelicerae are orange brown and in the male extended forward considerably. The abdominal dorsum is whitish with transverse rows of dark spots. Length of female 3.9 to 5.5 mm; of male 3.7 to 5 mm.

Virginia south to Florida and west to Arkansas and Texas; also California.

FAMILY ZORIDAE

This is a small family with a single genus and single species in our area. The spiders hunt in tall grass and bushes during daylight. The egg sac is guarded by the female, but unlike the clubionids no protective retreat is built.

Figure 573. *Zora pumila.*

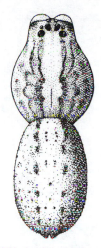

Figure 574. *Anahita animosa,* male.

Figure 573. *Zora pumila* (Hentz)

The general color is yellow, with markings in orange brown. There is a wide band extending back from each posterior lateral eye and a thin Y-shaped mark on the abdomen, as well as other spots. The legs are spotted, the spots being so close together on patellae and tibiae that these segments look quite brown. Tibiae I and II have eight pairs of ventral spines which are long and overlapping. Length of female 5 to 6 mm; of male 3.5 to 4 mm.

New England south to Alabama.

FAMILY CTENIDAE

These are wandering spiders hunting prey over foliage and the ground. They are limited to the southern part of the area. There are four genera in our region.

1a **Labium broader than long. Retromargin of cheliceral fang furrow with three teeth.** . . . . *Anahita* (**1 species**)

Figure 574. *Anahita animosa* (Walckenaer)

This species is also known under the name *punctulata* (Hentz). The carapace is orange brown with a lighter central band either side of which is a wavy band extending back from the posterior lateral eyes. The abdomen is yellow with a double longitudinal row of black spots and small spots irregularly distributed. The legs are orange with many black spots, are much longer in the male than female, and have five pairs of long overlapping spines under tibia I. Length of female 7 to 9 mm; of male 6 to 8 mm.

Southern and central states.

1b **Labium longer than broad. Retromargin of cheliceral fang furrow with four or five teeth** . . . . *Ctenus* (**3 species**)

Figure 575. *Ctenus hibernalis,* male.

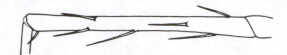

Figure 576. *Heteropoda,* tibia I, from in front, showing spines.

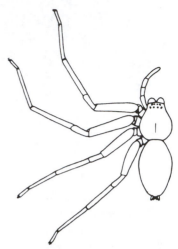

Figure 577. *Heteropoda venatoria.*

Figure 575. *Ctenus hibernalis* (Hentz)

The cephalothorax and legs are orange brown with a lighter median band on the carapace, set with white pubescence just behind the eye area. The abdomen is gray with a median yellow band that has serrated edges, and encloses dark spots. The chelicerae are quite heavy and geniculate. The legs are provided with many spines, there being five pairs under tibia I. Length of female 16 to 18 mm; of male 13 mm.

Alabama and Mississippi.

FAMILY SPARASSIDAE

These are the giant crab spiders, mostly tropical, and indigenous only to our warm regions. Some spider workers prefer the name Heteropodidae. There are three genera in our region.

1a Anterior median eyes smaller than the anterior laterals. Clypeus higher than the diameter of an anterior median eye. Tibia I with three or four pairs of ventral spines, the last pair distal and shorter (fig. 576) . . . . *Heteropoda* (1 species)

Figure 577. *Heteropoda venatoria* (Linnaeus)

The carapace is yellow to brown with black pubescence near the hind part. The abdomen is light tan with two or three indistinct longitudinal black lines. The marks are more conspicuous in males. Length of female about 23 mm; of male about 20 mm.

This spider occurs from Florida to Texas. It may occasionally be found in northern fruit stores on bunches of bananas having been accidentally shipped from its native habitats in Central America.

1b Anterior median eyes as large as or larger than the anterior laterals. Clypeus lower than the diameter of an anterior median eye (fig. 579). Tibia I with only two pairs of ventral spines, none at the distal end of the segment (fig. 578) . . . . *Olios* (5 species)

Figure 578. *Olios*, tibia I, from in front, showing spines.

Figure 579. *Olios*, face and chelicerae.

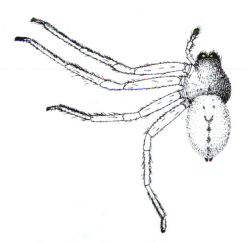

Figure 580. *Olios fasciculatus*, female.

Figure 580. *Olios fasciculatus* Simon

The carapace is orange with a sparse black pubescence. The abdomen is covered with gray hairs and there is a Y-shaped black mark extending the length of the dorsum. To the sides of this "Y" are several pairs of dark spots. Length of female 19 to 21 mm; of male 10 to 17 mm.

New Mexico and Utah west to California.

FAMILY SELENOPIDAE

The members of this family have very flat bodies and are found under stones, and flattened against rocks along the cracks of which they run very rapidly. They are easily recognized by the eye arrangement, having six eyes in the front row; the posterior median eyes have moved laterally and forward. In our region there is but a single genus, with four species.

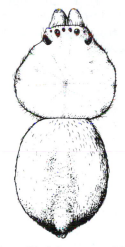

Figure 581. *Selenops actophilus.*

Figure 581. *Selenops actophilus* Chamberlin

The body and legs are a dirty yellow, with a mottling of gray spots. This makes for a perfect camouflage against the granite rocks over which they run. Length of female 11 to 12 mm; of male 9 to 9.5 mm.

Arizona and California.

FAMILY THOMISIDAE

Crab Spiders

The lateral eyes are either elevated alone or on conjoined tubercles. The cheliceral fang furrow is not very distinct and the margins are usually unarmed, or the promargin may have one or two teeth. Claw tufts are usually lacking or sparse. In most cases the body is depressed, being considerably wider than high. Laterigrade legs are quite characteristic.

These spiders are wanderers and secure their prey by stealth. They weave neither snares, retreats, molting nor hibernating nests. There are nine genera in our region.

1a **Clypeus (a) strongly sloping. Abdomen rather high and sloping upward toward the posterior end where there is a tubercle (b). (fig. 582)..................** ....*Tmarus* (5 species)

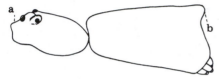

Figure 582. *Tmarus angulatus* from the side.

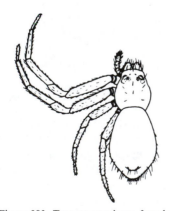

Figure 583. *Tmarus angulatus,* female.

Figure 583. *Tmarus angulatus* (Walckenaer)

The general color is brown mottled with yellow or white. The abdomen is indistinctly marked with chevrons on the posterior half. The size and height of the caudal tubercle vary in different individuals; it is smaller and lower in males. Length of female 4.5 to 7 mm; of male 3 to 5 mm.

Entire United States and southern Canada.

1b **Clypeus vertical. Abdomen flatter, broadly rounded behind and without a caudal tubercle...................2**

2a **Tubercles of lateral eyes confluent.....3**

2b **Tubercles of lateral eyes discrete, usually well separated 5**

3a **Anterior lateral eyes larger than the anterior medians. Carapace and abdomen spinose. Legs spinose, especially on the prolateral surface of femur I ....** ....*Misumenops* (20 species)

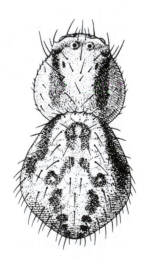

Figure 584. *Misumenops asperatus.*

Figure 584. *Misumenops asperatus* (Hentz)

The dorsal surface of the body and the legs are covered with short stiff hairs. The ground color is yellow or white with reddish markings. The front legs are marked with red annuli on tibiae and tarsi. Length of female 4.4 to 6 mm; of male 3 to 4 mm.

In grass and foliage.

Eastern United States and southern Canada to Arizona and north to Alberta.

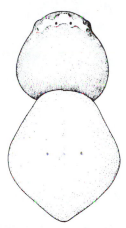

Figure 585. *Misumenops oblongus.*

Figure 585. *Misumenops oblongus* (Keyserling)

In the female the carapace has fewer spines than *asperatus*. The spines on leg I are unusually weak for a member of this genus. The carapace is pale greenish to white, with males often showing a red marginal stripe. The abdominal dorsum is pale green to silvery white, only occasionally margined in red. Length of female 4.9 to 6.2 mm; of male 1.5 to 2.6 mm.

Throughout the United States but more common in the south.

Misumenops deserti Schick
This species is quite variable in coloration, from light to dark brown, and in life often with tinges of red or green. There is a central white band on the carapace, widening in front where it reaches the clypeal margin. The male may be grayer, with the abdomen white or light greenish yellow and with large spots on either side of the median area. Length of female 6 to 6.5 mm; of male 3.5 to 4 mm.

Southwestern States.

Misumenops importunus (Keyserling)
Similar to *deserti,* but with the central white band on the carapace not extending to the clypeal margin and narrower in front. The abdomen is generally white to yellow in the female and brownish in the male. Length of female 7 mm; of male 3.5 mm.

Pacific Coast States.

Misumenops lepidus (Thorell)
The lateral bands on the carapace are gray to brown in females and reddish in males. The abdominal dorsum is white or yellow with darker markings on either side of the median area. These are often reddish in life. Length of female 5 mm; of male 3 mm.

Pacific Coast States.

3b Eyes of anterior row subequal in size. Carapace and abdomen devoid of strong spines. Legs with few or no dorsal or lateral spines 4

4a Carapace relatively flat. Clypeus with a distinct white carina (fig. 586) . . . .
. . . . ***Misumenoides*** **(1 species)**

Figure 586. *Misumenoides formosipes,* carapace from in front.

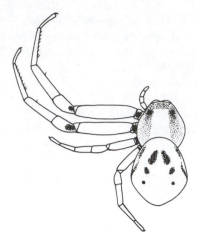

Figure 587. *Misumenoides formosipes*, female.

Figure 587. *Misumenoides formosipes* (Walckenaer)

This species is also known under the name *aleatorius* (Hentz). The carapace is creamy white to yellow or yellowish brown, with the sides slightly darker. In some there are broad red bands on the carapace and red spots on the legs. The abdomen may be unmarked, or there may be red or brown markings. In the male legs I and II are red or brown without lighter rings. Legs III and IV are yellow or white like the abdomen. Length of female 5 to 11.3 mm; of male 2.5 to 3.2 mm.

This species lives on plants and among flowers. Like the similar *Misumena vatia* it has been shown capable of changing its color to some extent.

Entire United States.

4b Carapace more convex. Clypeal carina absent (fig. 588). . . . .
. . . . *Misumena* **(1 species)**

Figure 588. *Misumena vatia*, carapace from in front.

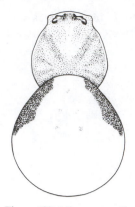

Figure 589. *Misumena vatia*.

Figure 589. *Misumena vatia* (Clerck) Flower Spider; Goldenrod Spider

In the female the carapace is white to yellow with the sides somewhat darker than the middle. The eye region is often tinged with red. The abdomen is usually the same color as the carapace, often without markings, but usually with a bright red band on either side, and occasionally with a median row of spots. The legs are light colored. In the male the carapace is dark reddish brown to red, with a creamy white spot in the center continuing to the eye area and clypeus. Legs I and II are reddish brown. Legs III and IV are yellow. On the abdomen is a pair of dorsal and a pair of lateral red bands, over a creamy white background.

Length of female 6 to 9 mm; of male 2.9 to 4 mm.

This is one of the most abundant of the "flower spiders" and like the similar *Misumenoides formosipes* has been shown capable of changing its color to some extent.

Entire United States and southern Canada.

5a Tibia I with two pairs of ventral spines. Median ocular area usually much longer than broad . . . .
. . . . *Oxyptila* **(20 species)**

GENUS OXYPTILA

The members of this genus are all gray or brown with lighter markings which make them resemble many species of *Xysticus,* though they are smaller in size, and have only two pairs of ventral spines under tibia I and II. They live on the ground under dead leaves.

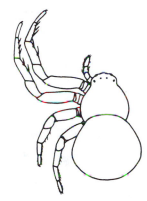

Figure 590. *Oxyptila conspurcata,* female.

Figure 590. *Oxyptila conspurcata* Thorell

Length of female 3 to 4 mm; of male 2.8 mm.

Wisconsin west to the Rockies and on to the Pacific Coast States. In Canada, Alberta east to Manitoba.

Oxyptila distans Dondale & Redner
Length of female 3 to 4 mm; of male 3 to 3.5 mm.

New England south to Georgia and west to Wisconsin. Also southern Canada.

Oxyptila monroensis Keyserling
Length of female 3 to 4 mm; of male 2.5 to 3 mm.

Pennsylvania south to Virginia and west to Kansas and Texas.

Oxyptila americana Banks
Length of female 3.5 to 4 mm; of male 3 to 3.5 mm.

New York and adjacent Canada west to Iowa and southwest to Texas.

5b Tibia I with three or more pairs of ventral spines. Median ocular area broader than long, or at least as broad as long 6

6a Carapace strongly convex (fig. 591). Claws on tarsus I with 7 to 12 teeth .... *Synema* (3 species)

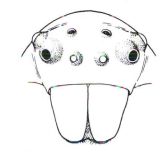

Figure 591. *Synema,* carapace from in front.

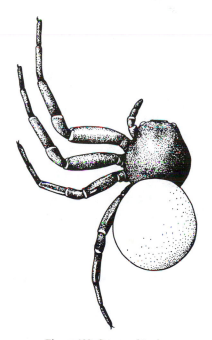

Figure 592. *Synema bicolor.*

Figure 592. *Synema bicolor* Keyserling

The carapace and legs are dark brown to black and the abdomen is light gray. Length of female 4.5 to 5 mm; of male 3 to 3.5 mm.

Found in tall grass and low bushes.

New England and adjacent Canada to Florida.

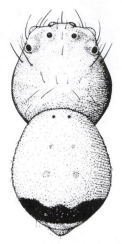

Figure 593. *Synema parvula*.

Figure 594. *Coriarachns*, carapace from in front.

Figure 595. *Coriarachne lenta*, female.

Figure 593. *Synema parvula* (Hentz)

This species is also known under the name *varians* (Walckenaer). The carapace and legs are yellowish orange, the eyes surrounded by white. The abdomen is yellow in front but has a broad transverse black band on the posterior fourth. Length of female 2 to 3 mm; of male 2.3 mm.

Found on flowers.

New Jersey and Illinois south to Florida and west to the Rockies.

6b Carapace less convex. Claws on tarsus I with fewer than 7 teeth. 8

7a Carapace very flat (fig. 594). Anterior eye row straight or slightly recurved. Cervical groove well indicated. . . . .
. . . . *Coriarachne* (4 species)

Figure 595. *Coriarachne lenta* (Walckenaer)

This species is also known under the name *versicolor* (Keyserling). The anterior row of eyes is recurved. The carapace is usually dark brown with numerous light spots and streaks which tend to fuse on the posterior declivity. The abdomen is gray with its margins darker, and usually with two or three pairs of light, alternating with darker, markings. Length of female 5.5 mm; of male 4.5 mm.

This species is common under loose bark, on fences, and under stones.

New England and adjacent Canada south to Florida and west to Arizona.

Coriarachne utahensis (Gertsch)

In general this species resembles *lenta* but averages somewhat larger in size. Also, the anterior row of eyes is more strongly recurved, the anterior legs are spotted, and the spots on the posterior declivity of the carapace are well separated. Length of female 4.5 to 9.9 mm; of male 4.1 to 6.3 mm.

New England west across the northern tier of States to the Rocky Mountain States, and to the Pacific Coast States. All of Canada and Alaska.

Coriarachne floridana Banks

The entire body is marked with brown, yellow, and white mottling. There is a pigmented stripe on the ventral surface of the legs. The spots on the carapace declivity of the female are contiguous, or nearly so. The anterior row of eyes is weakly recurved. Length of female 4.9 to 7.9 mm; of male 3.4 to 5.3 mm.

New York south to Florida and west to Louisiana and Arkansas.

Coriarachne brunneipes Banks

The carapace is flatter in this species than in the preceding three. The carapace and legs are of a uniform reddish brown color. The abdominal dorsum is mottled with brown, black, yellow, and white. The anterior row of eyes is straight. Length of female 6.3 to 11.2 mm; of male 3.8 to 6.3 mm.

Rocky Mountain States west to the Pacific Coast States. In Canada from Ontario to Alaska.

7b Carapace rather high (fig. 596). Anterior eye row moderately recurved. Cervical groove indistinct or absent. . . . .
. . . . *Xysticus* **(68 species)**

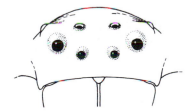

Figure 596. *Xysticus,* carapace from in front.

GENUS XYSTICUS

There are numerous species in this genus and most are various shades of brown or gray with white or yellow markings with the males usually darker. Often there is sexual dimorphism so that a great deal of confusion has resulted from the two sexes being matched up improperly.

These spiders live on and under loose bark, under leaves and stones of the forest floor, and on low plants.

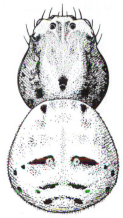

Figure 597. *Xysticus triguttatus.*

Figure 597. *Xysticus triguttatus* Keyserling

The carapace is yellowish brown in the female with a row of black marks along each side. The median lighter band is marked at the thoracic groove with a black spot. The eyes are situated on creamy white tubercles. The abdomen is white or light gray above with two small black spots in front and several transverse rows of spots behind. In the male the carapace is much darker. Length of female 4 to 6 mm; of male 3 to 5 mm.

New England and adjacent Canada south to Georgia and west to the Rockies.

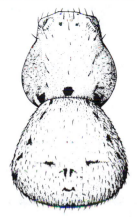

Figure 598. *Xysticus gulosus.*

Figure 598. *Xysticus gulosus* Keyserling

The pattern is similar to that of *triguttatus*. The general color is uniform grayish brown, darker in the male. There is an indistinct broad median pale band on the carapace, and the dorsum has a few small black spots. Femur IV has a distinct black spot at its distal end. Length of female 5 to 8 mm; of male 4 to 4.5 mm.

New England and adjacent Canada south to Georgia and west to California (but apparently not found in the Great Plains States and desert portions of the Southwest).

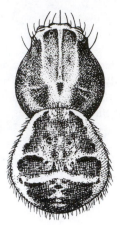

Figure 600. *Xysticus transversatus.*

Figure 600. *Xysticus transversatus* (Walckenaer)

This species is also known under the name *ferox* (Hentz). The pattern is somewhat similar to that of *elegans* but with a more conspicuous central light band on the carapace, and with the body as a whole more gray, not quite as brown. Length of female 6 to 7 mm; of male 5 to 6 mm.

New England south to Georgia and west to Texas and the Rockies. In Canada west to Alberta.

Figure 599. *Xysticus elegans*

Figure. 599. *Xysticus elegans* Keyserling

The carapace is brown, somewhat lighter in the middle. The abdominal dorsum shows three pairs of large brown spots, more distinct in males, alternating with light areas. The legs show a fine light line on the dorsal side, and are marked all over with small light brown spots. Length of female 8 to 10 mm; of male 6 to 7 mm.

New England and adjacent Canada south to Georgia and west to the Rockies.

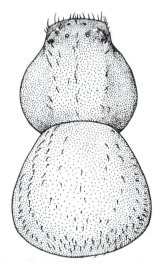

Figure 601. *Xysticus tumefactus.*

Figure 601. *Xysticus tumefactus* (Walckenaer)

This species is also known under the name *funestus* Keyserling. The general color is evenly light brownish yellow to rusty red, with very small light spots all over the body. Length of female 6 to 7 mm; of male 4 mm.

New England and adjacent Canada southwest to Texas, Oklahoma and the southern Rockies.

Xysticus locuples Keyserling
In both sexes the carapace is dark gray to brown to brick red. The abdomen is tinged with orange markings. Length of female 6.5 to 8.5 mm; of male 4.4 to 5.5 mm.

Montana south to New Mexico and west to the Pacific Coast States and British Columbia. In California, at least, this species has been collected in chapparal.

Xysticus montanensis Keyserling
The carapace has a median longitudinal light band narrowing behind to about half its width. The sides are brown to black, darker in males. The abdomen is brown to black with an indistinct pattern. Length of female 5.6 to 7 mm; of male 3.4 to 4.5 mm.

Montana south to Arizona and west to the Pacific Coast north to Alaska.

FAMILY PHILODROMIDAE

By many this is still considered a subfamily of the Thomisidae. But, in addition to those characters indicated in the key to families (page 52) there are others that seem sufficient to prompt separation. These include internal eye structure, chromosome constitution, and many characters of the developing spiderlings. There are eight genera in our region.

1a **Leg II at least twice as long as the others, which are subequal** . . . .
. . . . ***Ebo* (20 species)**

GENUS EBO

The members of this genus are easily recognized on the basis of leg II being much longer than the others. The clypeus is not quite as high as the length of the median ocular area. The anterior median eyes are largest, with the other eyes subequal, and with the posterior medians closer to the posterior laterals than to each other.

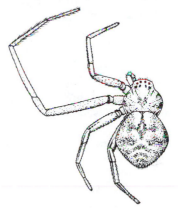

Figure 602. *Ebo latithorax,* female.

Figure 602. *Ebo latithorax* Keyserling

The general color is yellowish brown with darker markings on carapace and abdomen. The height of the clypeus is less than the distance between the anterior median eyes. The legs are devoid of dense scopulae on the tarsi, and the tibiae and metatarsi I and II lack long prolateral spines. The tibiae and metatarsi I, III, and IV are provided with a dorsal stripe along the length of the segments. Length of female 2.5 to 3 mm; of male 2 to 2.5 mm.

New England to Georgia and west to Nebraska and Texas.

Ebo pepinensis Gertsch
The carapace and legs are orange brown with dark spots. The abdomen is whitish to light gray with darker cardiac mark and lateral stripes. The clypeus, leg scopulae, and leg spines are similar to those of *latithorax.* But it differs from that species in lacking the dorsal

dark stripe on the tibiae and metatarsi I, III, and IV. Length of female 3.5 to 5.9 mm; of male 2.2 to 2.8 mm.

Kansas and Texas northwest to the northern Rockies and adjacent Canada, then west to the Pacific Coast States.

Ebo mexicanus Banks
The general color is white to yellow, with orange to brown along the sides of the carapace. The legs are spotted with brown. The abdomen has a gray cardiac mark and lateral stripes. The height of the clypeus is greater than the distance between the anterior median eyes. The tarsi are densely scopulate and the tibiae and metatarsi I and II have long prolateral spines. Length of female 3.4 to 4.4 mm; of male 2.9 to 3.6 mm.

Texas west to California.

Ebo parabolis Schick
Very similar to *mexicanus* with respect to coloration and size, height of clypeus, and leg spines. However, the dorsum is provided with patches of orange scales toward the sides and rear. Length of female 3.9 to 4.9 mm; of male 3.3 to 4.1 mm.

Colorado and New Mexico west to the Pacific Coast States.

Ebo californicus Gertsch
Similar to *mexicanus* in coloration, height of clypeus, and leg spines, but somewhat larger in size. Length of female 4 to 5.2 mm; of male 3.2 to 4.1 mm.

Wyoming south to Arizona and west to California and Oregon.

1b   **Leg II very little longer than I** 2

2a   **Posterior median eyes distinctly farther from each other than from the laterals** . . 3

2b   **Posterior eyes equidistant or with the medians farther from the laterals than from each other** 4

3a   **Posterior eyes in a slightly recurved line, the medians one and a half times as far from each other as from the laterals. Carapace about as wide as long or wider** . . . . *Philodromus* **(51 species)**

GENUS PHILODROMUS

Most of these spiders live on plants over which they can run very rapidly and their flat bodies enable them to get under cracks in bark. They are protectively colored and difficult to see except when they move. The cocoons are usually fastened to a leaf, twig or stone, and are guarded by the mother.

Figure 603. *Philodromus vulgaris.*

Figure 603. *Philodromus vulgaris* Hentz

For many years this species was confused with *pernix* Blackwall. The general color is gray and brown with dark spots and bands. There is a little light area at the posterior end of the abdomen. The legs are spotted, the spots being largest at the end of the segments. Length of female 4.5 to 8 mm; of male 4.5 to 6.3 mm.

New England to Florida and west to Wyoming. Also southern Canada.

Figure 604. *Philodromus marxi.*

Figure 604. *Philodromus marxi* Keyserling

This species is also known under the name *abbotii* Walckenaer. The general color is dark brown and milky white, with the male darker, orange to brown. Length of female 2.7 to 3.8 mm; of male 2.3 to 3 mm.

    New England to Florida and west to Texas and Nebraska.

Figure 605. *Philodromus placidus,* female.

Figure 605. *Philodromus placidus* Banks

In general appearance very similar to *marxi.* The legs ars yellow with a few brown markings that are most noticeable on III and IV. At about the middle of femora III and IV there is a dark line running across the dorsal surface of the segment, and from this another streak of pigment extending distally on the prolateral surface. The male is darker and the eyes are

ringed in white. Length of female 3.5 to 5 mm; of male 3 to 3.5 mm.

    New England and adjacent Canada south to Florida, west to Texas, then northwest through the Great Plains to Alaska.

Philodromus cespitum (Walckenaer)
This species in color and markings is intermediate between *vulgaris* and *imbecillus.* Length of female 4.5 to 6.1 mm; of male 4 to 5 mm.

    New England to New Jersey, west to the Rockies, and to the Pacific Coast States. In Canada across to Alaska.

Figure 606. *Philodromus imbecillus.*

Figure 606. *Philodromus imbecillus* Keyserling

The colors are brown and creamy yellow with a pattern as illustrated. Length of female 3.5 to 4.5 mm; of male 3 to 3.7 mm.

    New England south to Florida and west through the northern States to the Rockies. In Canada west to Alberta.

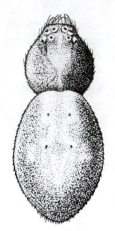

Figure 607. *Philodromus rufus.*

Figure 607. *Philodromus rufus* Walckenaer

The colors are reddish and yellow, with the eyes very conspicuously ringed in white. There are three subspecies. Length of female 2.7 to 4.5 mm; of male 2.5 to 3.5 mm.

New England southwest to Tennessee, west across the northern States to the Rockies and the Pacific Coast States. Also, virtually all of Canada to Alaska.

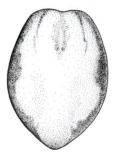

Figure 608. *Philodromus minutus,* female.

Figure 608. *Philodromus minutus* Banks

The carapace is mottled brown on the sides and yellowish along the broad median band. The abdomen has brown markings on a yellow ground. The legs are yellow with a few small

dark spots. The male is somewhat darker. Length of female 2.7 to 4.2 mm; of male 2.3 to 2.8 mm.

New England and adjacent Canada south to Georgia and west to Minnesota and Texas.

Philodromus spectabilis Keyserling
In some specimens the sides of the carapace are streaked with brown but with a central band that is lighter. Others show the lighter band only faintly. The abdominal dorsum is brown with a darker median band and lateral bands which converge toward the rear. Length of female 5.2 to 6.7 mm; of male 4.5 to 5.5 mm.

Montana south to New Mexico and west to the Pacific Coast States and British Columbia.

Philodromus histrio (Latreille)
The carapace is whitish to yellow or brown, mottled with gray spots. The legs are gray with brown spots. The abdominal dorsum is grayish green with a green or gray cardiac mark and a series of chevrons behind. Length of female 5.5 to 7 mm; of male 4.5 to 5.5 mm.

Michigan west across the northern States to the Rockies and west to the Pacific Coast States. In Canada Nova Scotia, then Manitoba west to British Columbia.

Philodromus alascensis Keyserling
Very similar to *histrio.* Length of female 5 to 6.5 mm; of male 4.5 to 5 mm.

Minnesota west across the northern States to the Rockies and west to the Pacific Coast States north to Alaska.

3b **Posterior eyes in a strongly recurved line, equidistant or almost so. Carapace slightly longer than wide. (The anterior lateral eyes are about as large as the anterior medians.)** . . . .
. . . . *Apollophanes* (6 species)

Figure 609. *Apollophanes* carapace.

Figure 609. *Apollophanes texanus* Banks

The ground color is yellow, with brownish spots. Length of female 5.4 to 8 mm; of male 5 mm.

Texas and the Rocky Mountain region west to California.

4a   Carapace almost (.9 to .95) as wide as long. Abdomen from one and a fourth to one and three-fourths as long as wide. (The anterior lateral eyes are generally larger than the anterior medians.) . . . . *Thanatus* (8 species)

GENUS THANATUS

In this genus the legs are almost all the same length, with leg II or IV longest, and the latter longer than I. The colors are brown or gray and the pattern is the same in all.

Figure 610. *Thanatus formicinus.*

Figure 610. *Thanatus formicinus* (Clerck)

Femur II is longer than the width of the carapace. In the female there are two prolateral spines on femur I. In the male the palpal tibia is provided with two or three dorsal spines and one prolateral spine. Length of female 6 to 8 mm; of male 5 to 6 mm.

Specimens have been swept from grass and bushes and collected from the trunks and branches of trees.

New England southwest to Alabama and west to the Pacific. Also virtually all of Canada to Alaska.

Thanatus vulgaris Simon
Similar to *formicinus* with respect to femur II. However, typically there are three prolateral spines on femur I, and the palpal tibia of the male has only one dorsal spine. Length of female 5 to 10 mm; of male 4.5 to 6 mm.

Ohio south to Georgia and west to Idaho, and also California.

Thanatus coloradensis Keyserling
Similar to *vulgaris* with respect to femur II and the spination of I. But the palpal tibia of the male has three or four dorsal spines. Length of female 8 mm; of male 6 mm.

Maine and adjacent Canada west through the northern States to the Rockies and the Pacific Coast States and to Alaska.

Thanatus striatus C.L. Koch
This species is much smaller than the preceding three. The length of femur II is less than the width of the carapace. There are only two prolateral spines on femur I and the palpal tibia in the male has one dorsal and two prolateral spines. Length of female 3.5 to 5 mm; of male 2.5 to 4 mm.

New England and adjacent Canada west across the northern States to the Pacific Coast States and to Alaska.

4b   Carapace not more than four-fifths as wide as long. Abdomen from two and a half to five times as long as wide . . . . *Tibellus* (6 species)

Figure 611. *Tibellus oblongu.*

Figure 611. *Tibellus oblongus* (Walckenaer)

The general color is light gray or yellow with darker longitudinal stripes. The abdomen is about three times as long as wide and at about one-third its length from the rear is a pair of small black spots. Leg II in the male is about five times, in the female about four times as long as the carapace. The legs are light without markings, and there are three pairs of ventral spines on tibia I. Length of female 7 to 9 mm; of male 6 to 8 mm.

Fairly common in bushes and tall grass and can be collected with a sweep net.

Throughout the United States and southern Canada.

atsoolleg*Tibellus duttoni* (Hentz)
Like *oblongus,* but slimmer, the abdomen being three and one-half to five times as long as wide; and tibia I with four pairs of ventral spines. Leg II in both sexes is five times as long as the carapace. Some specimens show two pairs of spots on the abdomen. Length of female 6 to 10 mm; of male 5 to 7 mm.

New England south to Florida and west to Texas and Minnesota.

atsooll*Tibellus chamberlini* Gertsch
Similar to *duttoni* in being slimmer than *oblongus,* and in having sometimes a second

pair of black spots on the abdomen, and in having four pairs of ventral spines on tibia I. In the female the abdomen is three and a third times as long as wide and leg II is not quite four and a half times as long as the carapace. In the male the abdomen is four times as long as wide, and leg II is five and a half times as long as the carapace. Length of female 9.6 to 11 mm; of male 7 to 8 mm.

Idaho south to Arizona and west to California.

FAMILY SALTICIDAE
Jumping Spiders

This is a very large family with 40 genera in our region and with the number of species increasing as one approaches the warmer portions of the continent. They are all easily recognized by the characteristic arrangement of the eyes, and have the keenest vision of all spiders. They make use of the visual sense to hunt their prey in broad daylight, and are common in sunny areas. In stalking prey they approach the latter slowly until a short distance away then make a sudden quick jump. Just before jumping the front legs (which are usually the heaviest) are extended forward for seizing the prey and an anchor line is made fast by the spinnerets. While no snare is built they all construct closely-woven retreats for molting, hibernating, and spending the night, under bark, between stones, or in rolled leaves. Often several spiders of the same species will build hibernating nests in close proximity.

1a Tibia plus patella III as long as or longer than tibia plus patella IV 2

1b Tibia plus patella III shorter than IV . . . 5

2a Ocular quadrangle as wide in front as behind, or wider in front 3

2b   Ocular quadrangle wider behind 4

3a   Anterior eye row noticeably wider than posterior. Sternum in front wider than the base of labium, which is as wide as long. Height of cephalothorax more than half its length. Leg III longer than I . . . . *Habrocestum* **(7 species)**

Figure 612. *Hobrocestum pulex.*

Figure 612. *Habrocestum pulex* (Hentz)

In the female the general color is grayish brown, with a lighter triangular stripe on the carapace and lighter irregular markings on the abdomen. The legs are light and dark spotted. In the male the cephalic part is black, while the rest of the carapace is orange brown to yellow. The abdomen is darker than in the female but similarly marked. In both sexes, the ocular quadrangle occupies about two-fifths of the length of the carapace and is narrower behind than in front. The legs are about equally stout, with III and IV longer than I and II. Length of female 4.5 to 5.5 mm; of male 4 to 4.5 mm.

This is a common spider actively hopping about on gray stones and dry leaves.

New England and adjacent Canada south to Florida and west to Louisiana and Nebraska.

3b   Anterior eye row barely or not at all wider than posterior eye row. Sternum in front narrower than base of labium which is longer than wide. Height of cephalothorax less than half its length. Leg III not longer than I . . . . *Plexippus* **(2 species)**

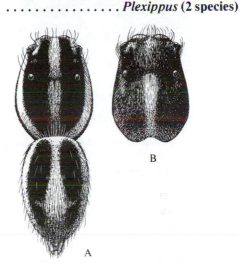

Figure 613. A, *Plexippus paykulli,* male; B, Carapace of female.

Figure 613. *Plexippus paykulli* (Audouin)

The cephalic region is black, the thoracic brown with a light median stripe. The abdomen is black with a yellow median line, and white lateral lines. In males the light median stripe extends farther forward on the carapace and there are white submarginal lines as well. Length of female 10 to 12 mm; of male 9.5 mm.

Georgia and Florida west to Texas.

4a   Leg I longer than III . . . . *Evarcha* **(2 species)[6]**

6. Some workers prefer to consider *Evarcha* and *Habronattus* as belonging in the genus *Pellenes* in the broad sense.

Figure 614. *Evarcha hoyi.*

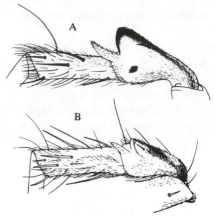

Figure 615. A, Patella and tibia III of male *Habronattus viridipes*. B, Patella and tibia III of male *Habronattus coronatus*.

Figure 614. *Evarcha hoyi* (G. & E. Peckham)

The general color is brown mixed with white and yellow scales and black hairs. There is always a light transverse band behind the cephalic part and a kind of herringbone pattern on the abdomen. The pattern is quite variable but many individuals will be found to resemble the figure. Length of female 4.6 to 6.3 mm; of male 4.3 to 5.5 mm.

This species is quite common in tall grass and bushes.

New England to Pennsylvania and west to the Pacific coast. In Canada from Ontario to British Columbia.

4b   Leg III longer than I . . . .
. . . . *Habronattus* **(57 species)**[7]

GENUS HABRONATTUS

The males of this genus often have peculiar modifications of form, color or ornament on leg I, or III (fig. 615), or both, which may be displayed before the female in courtship.

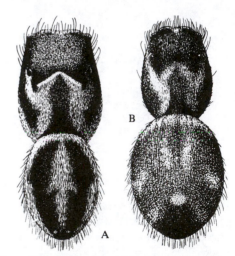

Figure 616. *Habronattus viridipes.* A, Male; B, Female.

Figure 616. *Habronattus viridipes* (Hentz)

In the male the cephalic area is orange brown, the thoracic area is black on the sides and on top with white between. The abdomen has a median and two lateral white bands alternating with two black or dark brown bands. The central white band is usually enlarged and notched in the middle, but often this is not the case, so

7. Some workers prefer to consider *Evarcha* and *Habronattus* as belonging in the genus *Pellenes* in the broad sense.

that the abdominal pattern is similar to that of the female.

Leg II in life has a light green color (usually fading to yellow after some time in alcohol). Leg I has a thick fringe of white hairs on the femur, a thin short fringe on the patella, and a longer one on the tibia. Leg III has the patella widened at the distal end, with a conspicuous black spot on its prolateral face, and a stout spur arising from close to the base and extending over the tibia.

The color of the female is dull orange brown, the light areas being tan rather than white. Length of female 5.5 to 7 mm; of male 4.3 to 5.5 mm.

Collected from open woodland, from under stones and along paths.

Eastern United States and adjacent Canada to Wisconsin.

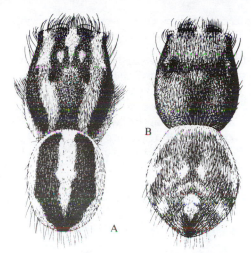

Figure 618. *Habronattus agilis.* A, Male; B, Female.

Figure 618. *Habronattus agilis* (Banks)

The male is strikingly marked with three black and four white longitudinal stripes on the carapace. The abdomen has white stripes and spots. Leg I is ornamented with long fringes of hair on femur, patella and tibia. The female is gray with less distinct white spots on the abdomen. Length of female 5.2 to 6.4 mm; of male 5 to 5.5 mm.

New England to Florida and west to New Mexico.

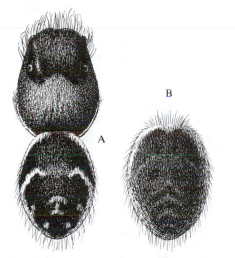

Figure 617. *Habronattus borealis.* A, Male; B, Abdomen of female.

Figure 617. *Habronattus borealis* (Banks)

The male has the cephalothorax all black, and the abdomen black with diagonal white spots. The female is light gray and brown with several light spots in the median line behind. Length of female 5 to 6 mm; of male 4.5 to 6 mm.

New England and adjacent Canada south to North Carolina and west to Washington.

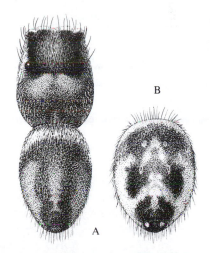

Figure 619. *Habronattus decorus.* A, Male; B, Abdomen of female.

Figure 619. *Habronattus decorus* (Blackwall)

The male is covered with white hairs, the posterior two-thirds of the abdomen showing pink with an iridescent sheen, through which is visible a dark transverse band near the front and two broad longitudinal stripes extending back and converging to meet at the spinnerets. The female is yellow to brown with the bands on the abdomen often broken into several large black spots.

Length of female 5 to 6 mm; of male 4 to 5 mm.

New England and adjacent Canada to Florida and west to Colorado.

Habronattus coronatus (Hentz)

In the male the cephalic part is covered with orange hairs. On the thoracic part is a pair of white longitudinal bands extending back from the rear eyes. The abdomen is grayish yellow with a white basal band, a central spot and several pairs of lateral light spots. Leg III has the distal end of the femur rounded into a shiny boss above, patella III is only slightly dilated distally, and has a short spur at its end. Tibia III is green in color, the other segments being yellow to brown. Length of female 5.5 mm; of male 4.3 to 4.7 mm.

More common in the southern states but known from New England west to California.

5a Spiders decidedly ant-like in form, with the body narrow, and the pedicel clearly visible from above **6**

5b Spiders not ant-like in form **8**

6a Posterior portion of cephalothorax narrowed and with its sides more or less parallel, thus adding to the apparent length of the pedicel . . . .
. . . . *Synemosyna* **(2 species)**

Figure 620. *Synemosyna lunata*.

Figure 620. *Synemosyna lunata* (Walckenaer)

This species is also known under the name *formica* Hentz. The cephalic part is separated from the thoracic by a deep groove, and there is a depression near the front of the abdomen as well. Seen from the side this all adds to the ant-like appearance of the spider. In both sexes there is a dorsal abdominal scutum, extending to the depression in females, and covering two-thirds the length of the dorsum in males. The cephalothorax and anterior portion of the abdomen are brown. At the depression there is a white line and behind this the abdomen is black. The legs are rather thin, the first not heavier than the others. Length of female 4.7 to 5.7 mm; of male 4.2 to 5 mm.

This species occurs on bushes and tall grass and has never been seen to jump. Instead it moves about like an ant.

New England and adjacent Canada south to Florida and west to Wisconsin.

6b Thoracic part not narrowed, and not parallel-sided in front of pedicel **7**

7a Carapace long and narrow, the width very little more than half the length. Female

with tibia and tarsus of pedipalp swollen (fig. 622). (Cervical groove [a] well marked, with the head region much higher than the steeply sloping thoracic portion (fig. 621)..................
.................*Sarinda* (2 species)

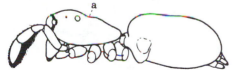

Figure 621. *Sarinda hentzi,* female from the side.

Figure 622. *Sarinda hentzi,* female.

New England to Florida and west to Texas and Kansas.

7b Width of carapace almost two-thirds the length. Female with pedipalp of the usual type, the tibia and tarsus not swollen....
..............*Peckhamia* (6 species)

Figure 623. *Peckhamia picata* from the side.

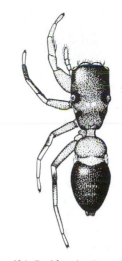

Figure 624. *Peckhamia picata,* female.

Figure 622. *Sarinda hentzi* (Banks)

In color and general appearance this spider resembles an orange brown ant. There is an indistinct light mark across the middle of the carapace and another on the abdomen. Length about 5 to 7 mm in both sexes.

On the abdomen is a dorsal sclerite, present in both sexes. It does not cover the entire width of the dorsum, but in females covers one-third and in males two-fifths the length. The legs are moderately long and thin for a member of this family. The female is peculiar among spiders in having the tarsus and tibia of the palp much swollen, after the fashion of immature males (fig. 622, b).

Fig. 624. *Peckhamia picata* (Hentz)

The carapace shows a conspicuous constriction behind the rear eyes (fig. 623). The general color is reddish brown, somewhat darker and with violet reflections in the ocular area, and glistening black on the posterior half of the abdomen. There is a pair of white spots between the rear eyes, and another pair at the sides of the abdominal constriction. In both sexes the dorsum is completely covered by a thick shiny scutum which extends down on the sides. Length of female 3.5 to 5 mm; of male 2.8 to 4 mm.

New England and adjacent Canada south to Florida and west to Texas and Nebraska.

8a No tooth on the retromargin of the cheliceral fang furrow . . . . *Sitticus* (12 species)

Figure 625. *Sitticus palustris.*

Figure 625. *Sitticus palustris* (G. & E. Peckham)

The general color is light brown with yellow markings in the female, and dark brown with white markings in the male. Length of female 5 to 6 mm; of male 3.5 to 5 mm.

New England to the Pacific Coast States. In Canada from Ontario west to the Northwest Territory and Manitoba.

8b At least a single tooth present 9

9a Tibia I without ventral spines . . . . *Salticus* (4 species)

Figure 626. *Salticus scenicus.*

Figure 626. *Salticus scenicus* (Clerck) Zebra Spider

The general color is gray with white markings, but often there are brown to reddish scales mixed in with the gray, and usually there are iridescent scales in the eye region. The chelicerae of the male are considerably elongated and extended almost horizontally forward, with the fangs long and sinuate. Length of female 4.3 to 6.4 mm; of male 4 to 5.5 mm.

This is a very common species on fences and outside walls of buildings. It frequently strays indoors.

Throughout the northern United States and southern Canada.

9b Tibia I with at least one ventral spine . . 10

10a Tibia I with four pairs of ventral spines. (Sternum in front narrower than base of labium) 11

10b Tibia I with fewer than four pairs of ventral spines. 12

11a Height of cephalothorax about two-thirds the greatest width. Tooth (a) on retro-

margin of cheliceral fang furrow with two cusps (fig. 627) . . . .
. . . . *Maevia* (3 species)

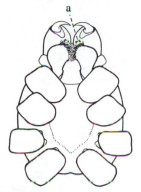

Figure 627. *Maevia inclemens*, cephalothorax from below.

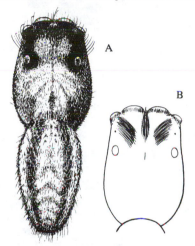

Figure 628. *Maevia inclemens*. A, Female; B, Carapace of male.

Figure 628. *Maevia inclemens* (Walckenaer)

This species is also known under the name *vittata* (Hentz). In one variety of male the body is black and there are three tufts of hairs on the cephalic part. The other variety is more like the female with red, black, and white markings on a gray ground. The female has the abdomen somewhat lighter to yellowish with chevrons on the posterior half. Length of female 6.5 to 10 mm; of male 4.8 to 7 mm.

New England and adjacent Canada to Florida and west to Texas and Wisconsin.

11b Height of cephalothorax only about half the greatest width. Tooth (a) on retro-margin of cheliceral fang furrow a simple one (fig. 629) . . . .
. . . . *Marpissa* (10 species)

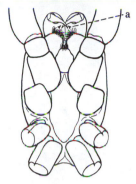

Figure 629. *Marpissa pikei*, cephalothorax from below.

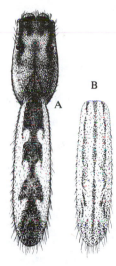

Figure 630. *Marpissa pikei*. A, Male; B, Abdomen of female.

Figure 630. *Marpissa pikei* (G. & E. Peckham)

The general color is light gray with brown markings. In the male there is a broad dark band running the length of the abdomen, and bordered each side by a narrow light band. In the female the central dark band is supplanted by three thin broken lines or rows of spots.

Length of female 6.5 to 9.5 mm; of male 6 to 8.2 mm.

New England and adjacent Canada to Florida and west to Nebraska and Arizona.

Figure 632. *Synageles noxiosa*, female from the side.

Figure 632. *Synageles noxiosa*, female from the side.

Figure 631. *Marpissa lineata*.

Figure 633. *Synageles noxiosa*, female.

Figure 631. *Marpissa lineata* (C.L. Koch)

On the abdomen are four white lines, the lateral two being so far down on the sides that only their anterior portions are visible from above. The legs in the female are brown, and in the male yellow, except for tibia I which is dark and hence in striking contrast to the other segments. Length of female 4 to 5.3 mm; of male 3 to 4 mm.

New England and adjacent Canada south to Georgia and west to the Mississippi River.

Figures 632 and 633. *Synageles noxiosa* (Hentz)

The general color is dull brown to gray with white or yellow markings. On the abdomen are two light transverse bands, the first at about the level of the shallow constriction. Between the two light stripes is a pair of light spots and on the whole this area is lighter than the rest of the dorsum. In both sexes a shiny dorsal scutum is present, occupying the entire dorsum in the male but only about the anterior quarter of the length in the female. Length of female 3 to 3.5 mm; of male 2 to 2.4 mm.

Found on fences and tree trunks.

Massachusetts west to Minnesota south to Florida and across the southern states to California.

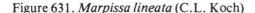

12a **Tibia I with at least four ventral spines** .. · **16**

12b **Tibia I with one, two, or three ventral spines** · **13**

13a **Ocular quadrangle longer than wide. Tooth on retromargin of cheliceral fang furrow bicuspid (as in fig. 627)** · *Synageles* (**14 species**)

13b **Ocular quadrangle wider than long. Tooth on retromargin of cheliceral fang furrow either simple (as in fig. 629) or else more than one tooth present** · · · · · · · **14**

14a Small eyes, of second row, midway between first and third rows. [With only one ventral and one prolateral spine on tibia I. Abdomen depressed, overhanging the cephalothorax (fig. 634), and with three to five transverse white bands.] . . . . *Ballus* (1 species)

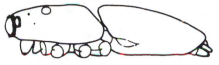

Figure 634. *Ballus*, lateral aspect.

Figure 635. *Ballus youngii*.

Figures 634 and 635. *Ballus youngii* G. & E. Peckham

The ocular area is black, the thoracic part dark brown, and the abdomen brownish with white scales arranged in three to five transverse narrow lines. The first line may be incomplete so as to appear as a pair of spots on the anterior half. Length of female 2.8 mm; of male 2.3 mm.

New England and adjacent Canada south to Florida and west to Wisconsin.

14b Small eyes closer to first row than to third 15

15a Width of cephalothorax seven-eighths the length, and height over half the length. Small eyes twice as far from the posterior row as from the anterior. Thoracic declivity steep and beginning closer to the posterior eyes than are the small eyes from the posterior eyes . . . . *Agassa* (1 species)

Figure 636. *Agassa cerulea*.

Figure 636. *Agassa cerulea* (Walckenaer)

This species is also known under the name *cyanea* (Hentz). The whole body is covered with iridescent scales which give it a green to purplish color, sometimes coppery brown. Leg I is much the heaviest. Length of female 3.3 to 4.6 mm; of male 3.1 to 4 mm.

New England to Florida and west to New Mexico.

15b Width of cephalothorax only five-sixths the length, and height only half the length. Small eyes just a little closer to the anterior row than to the posterior. Thoracic declivity not so steep and beginning behind the posterior eyes a distance about equal to that of small eyes in front of posterior eyes . . . . *Sassacus* (2 species)

Figure 637. *Sassacus papenhoei*.

Figure 638. *Tutelina*, cephalothorax of female from below.

Figure 637. *Sassacus papenhoei* G. & E. Peckham

Similar to *Agassa cerulea* in having the body covered by iridescent scales and leg I heaviest. There is a white marginal stripe each side of the carapace and on the abdomen a white basal band which is continued along the sides in the female. Length of female 4.4 to 5.5 mm; of male 2.8 to 4.7 mm.

Tennessee and Ohio west to the Pacific States and British Columbia.

16a Sternum in front narrower than the base of the labium 17

16b Sternum in front as wide as, or wider than, the base of the labium 20

17a Leg I not much if any thicker than the others. Labium wider than long (fig. 638) . . . . *Tutelina* (2 species)

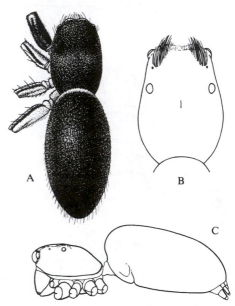

Figure 639. *Tutelina elegans*. A, Female; B, Carapace of male; C, Lateral aspect of body.

Figure 639. *Tutelina elegans* (Hentz)

In the male the ocular region is black, the thoracic part and abdomen brown, with the whole body covered by greenish scales making it brilliantly iridescent. There are conspicuous long tufts of hairs on the front. The abdomen is without spots or bands. The legs are pale with a black line above, most conspicuous on tibia I. On the distal third of the latter is a fringe of black hairs and a conspicuous black spot. The female is similar except that the cephalic tufts are lacking, the abdomen has a white basal band, and femur I is nearly all black. Moreover, the fringe and black spot are lacking from tibia I. Length of female 5.5 to 7 mm; of male 4 to 4.5 mm.

This species (and also the closely related *similis*) are quite like members of the genus *Icius*. They differ from *Icius* in that the height of the carapace is almost four-fifths the width and almost half the length; the carapace is slightly narrower; the ocular area occupies two-fifths the carapace length; the sternum is relatively narrower and its anterior truncature is slightly narrower than the base of the labium; the legs are lineate; and the endites of the male are drawn out apically to a point or hook. (Compare with Figure 666).

New England and adjacent Canada south to Florida, west to Washington.

17b Leg I much thicker than the others. Labium longer than wide 18

18a Tooth on retromargin of cheliceral fang furrow with two or more cusps (at least in females). At least legs II, III and IV white and translucent . . . .
. . . . *Hentzia* (4 species)

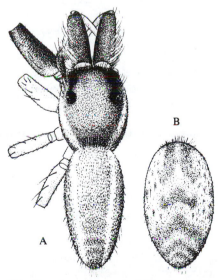

Figure 640. *Hentzia ambigua*. A, Male; B, Abdomen of female.

Figure 640. *Hentzia ambigua* (Walckenaer)

This species is also known under the name *palmarum* (Hentz). In the female the body is covered with light gray or white scales and on the middle of the dorsum is a row of brown spots and chevrons, often indistinct, and with oblique rows of small spots at the sides. The legs are translucent white to yellow. In the male the front legs are dark brown, except for the tarsi which are light. The other legs are as in the female. The carapace and abdomen are brown with coppery iridescence and a pair of lateral white lines. In many but not all males the chelicerae are elongated and project forward considerably. Length of female 4.7 to 6 mm; of male 3.7 to 5.5 mm.

New England and adjacent Canada south to Florida and west to Oklahoma and Nebraska.

18b Tooth simple. Legs II, III and IV with dark markings 19

19a Three pairs of spines under tibia I. Posterior declivity occupying about the posterior fourth of the carapace. Small eyes, of the second row, closer to the first than to the third row . . . .
. . . . *Menemerus* (2 species)

Figure 641. *Menemerus bivittatus*, male.

Figure 641. *Menemerus bivitattus* (Dufour)

The carapace is relatively flat, being only about three-tenths as high as long. The eye area is black, and the thoracic area is chestnut brown. There is a thin marginal line of white scales. The legs are yellow to light brown in the female and dark brown in the male. The abdominal dorsum shows a median gray band and a similar gray band on each side, with a lighter band each side of the middle. In the male there is a brownish scutum occupying about half the length and slightly less than half the width of the dorsum. Length of female 7 to 10.2 mm; of male 5.5 to 7.3 mm.

West Virginia south to Florida and west across the southern States to California.

19b Two pairs of spines under tibia I. Posterior declivity limited to the posterior fifth or sixth of the carapace. Small eyes of the second row midway between the first and third rows . . . . *Metacyrba* **(6 species)**

Figure 642. *Metacyrba taeniola.*

Figure 642. *Metacyrba taeniola* (Hentz)

The carapace is iridescent mahogany brown on the thoracic part and black on the ocular quadrangle. The posterior declivity is limited to the posterior fifth of the carapace. The abdomen is gray with two rows of whitish yellow narrow spots. Leg I is much thicker than the others and has the femur flattened. Length of female 6 to 7 mm; of male 5 to 6 mm.

Delaware south to Florida and west to California.

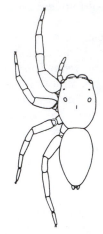

Figure 643. *Metacyrba.*

Figure 644. *Metacyrba undata,* abdomen.

Figures 643 and 644. *Metacyrba undata* (De Geer)

The cephalothorax is about four-fifths as wide as long, with the sides curved outward, and the rear eyes far removed from the lateral edges (fig. 643). The posterior declivity is restricted to the posterior sixth of the carapace. Both the cephalothorax and abdomen are much flattened and hairy. Iridescent scales are lacking. The general color is gray, the body being covered with an admixture of long, white, gray, and dull reddish hairs. In both sexes there are "eyebrow" hairs, those in the female being red, in the male white. On the abdominal dorsum is a pale yellow median stripe, within which is a darker line. Length of female 10 to 13 mm; of male 8.5 to 9.5 mm.

This species has been taken from under loose bark, on fences, and on the outside walls of buildings.

Eastern States and adjacent Canada to Texas and Wisconsin.

20a Small eyes, of the second row, closer to the third row than to the first row 21

20b Small eyes either equidistant between the first and third rows, or closer to the first row 22

21a Metatarsi longer than tarsi. Ocular quadrangle occupying two-fifths the length of the carapace. Height of carapace two-thirds its length .... *Corythalia* (4 species)

Figure 645. *Corythalia aurata*, female.

Figure 645. *Corythalia aurata* (Hentz)

The carapace is black in the cephalic region and chestnut brown in the thoracic part. There are white scales in a band along each lateral edge and in an elongated spot behind each rear eye. The abdominal dorsum shows a white line on each side in front and two pairs of large black spots behind. There is little difference between the sexes. Length of female 5.3 to 5.8 mm; of male 5 to 5.2 mm.

South Carolina south to Florida and west to Texas.

21b Metatarsi shorter than tarsi. Ocular quadrangle hardly one-third the length of the carapace. Height of carapace about two-fifths its length .... *Phlegra* (1 species)

Figure 646. *Phlegra fasciata*.

Figure 646. *Phlegra fasciata* (Hahn)

The ocular quadrangle occupies one-third the length of the carapace. Tibia I has three pairs of ventral spines. There is a pattern of two white lines on the carapace, and three on the abdomen alternating with broad brown stripes. The male is darker with brick red hairs in the eye region. Length of female 7 to 8 mm; of male 6 to 7 mm.

New England to Florida and west to Texas and Kansas.

22a Small eyes closer to the first than to the third row 23

22b Small eyes equidistant between the first and third rows 26

23a Ocular quadrangle more than half the length of the carapace . . . .
. . . . *Zygoballus* **(4 species)**

GENUS ZYGOBALLUS

The cephalothorax is widest at the rear eyes behind the middle of its length, and exhibits a steep declivity just beyond this. In the males the chelicerae are powerfully developed and provided on the lower surface near the lateral edge with a heavy hammer-like process (fig. 647).

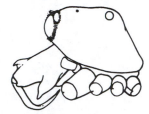

Figure 647. *Zygoballus,* cephalothorax of male from side.

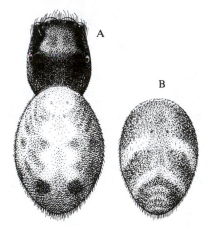

Figure 648. *Zygoballus bettini.* A, Female; B, Abdomen of male.

Figure 648. *Zygoballus bettini* G. & E. Peckham

In the female the carapace is dark brown, the abdomen lighter with a white basal transverse band and paired white spots behind. The male is darker, with red and green iridescence on the cephalic part. The abdomen has two pairs of white spots parallel to each other. Length of female 4.3 to 6 mm; of male 3 to 4 mm.

New England and adjacent Canada to Florida and west to Texas and Nebraska.

Figure 649. *Zygoballus sexpunctatus* male.

Figure 649. *Zygoballus sexpunctatus* (Hentz)

Similar to *bettini* but with a spot of white scales in the middle of the carapace and smaller, not parallel, white spots on the abdomen. Length of female 3.5 to 4.5 mm; of male 3 to 3.5 mm.

New Jersey to Florida and west to Texas.

23b Ocular quadrangle less than half the length of the carapace 24

24a Tibia I with two pairs of ventral spines. Width of carapace not over three-fourths the length (Retromargin of cheliceral fang furrow with three teeth). . . . .
. . . . *Thiodina* **(2 species)**

Figure 650. *Thiodina iniquies.*

Figure 650. *Thiodina iniquies* (Walckenaer)

This species is also known under the name *sylvana* (Hentz). The carapace is high and rounded with spots of white scales either side on the thoracic part. The abdomen has two white stripes bordered by small black spots. Length of female 8 to 10 mm; of male 7 to 9 mm.

    North Carolina south to Florida and west to California.

24b **Tibia I with three pairs of ventral spines. Width of carapace at least four-fifths the length** **25**

25a **Cephalothorax seen from above with sides strongly curved, the greatest width about seven-eighths the length. Small eyes twice as far from posterior row as from the anterior laterals (fig. 651) and the posterior row about one and a quarter times as wide as the first row. (Males often with tufts of hair in the eye region)** . . . . *Phidippus* **(50 species)**

Figure 651. *Phidippus*, carapace from side to show position of small eyes.

GENUS PHIDIPPUS

This genus includes our heaviest and hairiest jumping spiders. Many of the males have "eyebrow" tufts of hairs (fig. 652) and most have the chelicerae, at least in part, iridescent. Usually there is on the abdomen a distinct pattern including a light transverse basal band, light side bands and paired white spots above. The relative size of the spots varies and those of the second pair are often fused to form a central triangle. The sexes are often differently colored.

Figure 652. *Phidippus*, face and chelicerae.

Figure 653. *Phidippus audax.*

Figure 653. *Phidippus audax* (Hentz)

Mostly black with a white basal band on the abdomen and several spots behind, of which the central one is largest. While usually white, some or all of these spots may be yellow or orange, especially in young individuals. Some specimens, especially from the Southern States, have a white band on either side of the carapace, and are often larger. These have been known under the name *variegatus*. Length of female 8 to 15 mm; of male 6 to 13 mm.

This is an extremely common spider usually found running about over tree trunks and under stones and boards.

Atlantic coast states and adjacent Canada west to California.

Figure 655. *Phidippus texanus*.

Figure 655. *Phidippus texanus* Banks

As in *purpuratus* the carapace has a thick coating of gray hairs. The abdomen shows paired spots too, but there is a central longitudinal white band for most of its length. Length of female 12 to 13 mm; of male 8 to 9 mm.

Missouri west to Colorado and south through New Mexico, Texas and Arkansas.

Figure 654. *Phidippus purpuratus*.

Figure 654. *Phidippus purpuratus* Keyserling

The cephalothorax has a thick coating of gray hairs. The abdomen has four pairs of white spots, and on either side is a light band, white in females and yellowish in males. Length of female 12 to 15 mm; of male 10 to 12 mm.

This species may be found on and under stones and boards on the ground.

New England and adjacent Canada west to the Rockies.

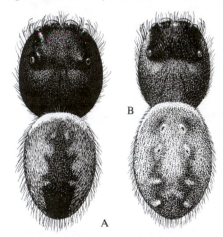

Figure 656. *Phidippus rimator*. A, Male; B, Female.

Figure 656. *Phidippus rimator* (Walckenaer)

This species is also known under the name *clarus* Keyserling. The female is yellowish orange to brown, with black and white markings. In the male the cephalothorax is black and the abdomen has a broad black band down the middle, with a white basal band and reddish lateral stripes. Between the red and black areas are often three pairs of white spots. Length of female 8 to 10 mm; of male 5 to 7 mm.

This species is found in tall grass and in bushes. It is the most common and the most widely distributed member of the genus.

Throughout the United States and southern Canada.

Figure 658. *Phidippus regius.*

Figure 658. *Phidippus regius* C.L. Koch

This species is somewhat similar in its markings to *otiosus*. The carapace has a larger area covered with white hairs. However, only a narrow area on each side where the eyes are, lack the white hairs. The abdominal dorsum varies from an orange brown to a brownish black ground color, the latter especially in the male. There are lighter spots as illustrated, the spots toward the rear being much smaller than the corresponding light areas that are present in *otiosus*. Length of female 13 to 19 mm; of male 9 mm.

North Carolina south to Florida (where it is much more common than *otiosus*) and west across the Gulf States to Texas.

Phidippus apacheanus Chamberlin and Gertsch The entire dorsal aspect of carapace and abdomen are a bright yellowish-orange. Length of female 11 to 12 mm; of male 9 to 11 mm.

Florida west through the southern tier of States to California.

Phidippus cardinalis McCook The entire dorsal aspect of carapace and abdomen are bright red. Length of female 9 mm; of male 8 mm.

Figure 657. *Phidippus otiosus.*

Figure 657. *Phidippus otiosus* (Hentz)

The carapace is dark brown with a thin covering of whitish hairs on the sides. The abdominal dorsum has the brown ground color but with lighter spots of orange hairs as illustrated. These include on either side of the middle a broad area, which in *regius* is reduced to a small spot. The chelicerae, even in females, may show an iridescence not only of the usual green color, but sometimes purplish to pinkish. Length of female 10 to 15 mm; of male 9 to 11 mm.

Maryland south to Florida and across the Gulf States to Texas.

New England south to Florida and west to Texas.

Figure 659. *Phidippus johnsoni.*

Figure 659. *Phidippus johnsoni* G. & E. Peckham

This species is the one that has been known for many years, especially in the literature on envenomation, under the name *formosus.* There have been numerous reports of its having bitten people, particularly in southern California, but none of the cases were serious.

The carapace is black in both sexes. In males the abdominal dorsum is all red, but in females this is seldom the case; usually the red appears on either side of a median black band on the posterior two-thirds of the dorsum. Length of female 8 to 13 mm; of male 7.2 to 11 mm. Specimens from the more northern parts of its range sometimes show white spots on the rear portion of the dorsum.

North Dakota and adjacent Canada south to Texas and west to the Pacific Coast States and British Columbia.

Figure 660. *Phidippus opifex,* male.

Figure 660. *Phidippus opifex* (McCook)

The carapace is concolorous reddish to mahogany brown in the male, and somewhat lighter with more orange, and with a gray pubescence in the female. The abdominal dorsum is orange with black markings as illustrated. The legs are mahogany brown. Length of female 13.7 to 15 mm; of male 9 to 11 mm.

Colorado and New Mexico west to California.

25b Cephalothorax with sides not quite so strongly curved, the greatest width about four-fifths the length. Small eyes not quite so near the anterior row as in *Phidippus,* and the third row is only about one and a tenth times as wide as the first row. (Males with chelicerae in many cases powerfully developed, projecting forward somewhat and with the fang sinuate) . . . .
. . . . *Eris* (6 species)

GENUS ERIS

This group is intermediate between *Phidippus* and *Metaphidippus.* Its members are generally larger and more hairy than the latter, but smaller and less hairy than the former. In the males leg I is fringed (fig. 661) and the chelicerae are powerfully developed (fig. 662), often extending forward slightly, and with the fangs sinuate.

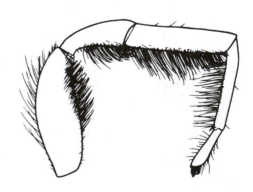

Figure 661. *Eris,* leg I of male showing fringes.

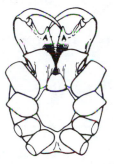

Figure 662. *Eris,* cephalothorax of male from below showing powerfully developed chelicerae.

Figure 664. *Eris aurantia,* female.

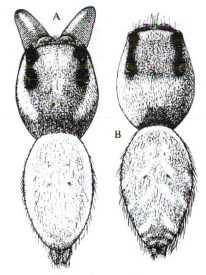

Figure 663. *Eris marginata.* A, Male; B, Female.

Figure 663. *Eris marginata* (Walckenaer)

The male has a bronze-brown body with a pair of white longitudinal bands along the sides of the dark cephalothorax and an encircling white band on the lighter abdomen. The female has the carapace lighter and without the white bands. On the abdomen there is a basal white transverse band and several pairs of somewhat oblique white spots. Length of female 6 to 8 mm; of male 4.7 to 6.7 mm.

This is a common spider in shrubbery and tall grass.

Throughout the United States and southern Canada.

Figure 664. *Eris aurantia* (Lucas)

This species is exceedingly variable in its markings, and most of the body is covered with iridescent scales so that the colors appear different depending upon the angle of reflected light. In the female the carapace is reddish brown on the thoracic part and black on the ocular quadrangle, and a band of white scales extends from the anterior lateral eyes around to the posterior declivity on each side. The abdomen is orange brown with a white, or bright orange, basal band, three or four pairs of white spots, and two pairs of orange spots. In the male the ground color is mahogany to black, and the orange spots are less conspicuous. Length of female 8 to 12 mm; of male 7 to 10 mm.

Delaware and Illinois south to Florida and west to Arizona.

26a Ocular quadrangle almost two-thirds the length of carapace . . . . *Neon* (6 species)

Figure 665. *Neon nellii.*

Figure 665. *Neon nellii* G. & E. Peckham

The general color is brownish gray, darker in the eye region. The abdomen is gray with a yellowish herringbone pattern of chevrons. Length of female 2.3 to 3 mm; of male 2 to 2.5 mm.

New England south to Florida and west to the Rockies. In Canada from Ontario west to Saskatchewan.

26b Ocular quadrangle occupying not more than half the length of carapace 27

27a Ocular quadrangle occupying one-half the length of carapace . . . .
. . . . *Icius* (11 species)

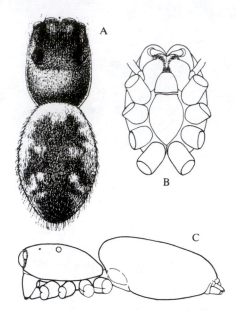

Figure 666. *Icius hartii.* A, Male; B, Cephalothorax from below; C, Body from the side.

Figure 666. *Icius hartii* Emerton

There is a light encircling band on the front and sides of the abdomen. The carapace is brown with a grayish pubescence and the abdomen is dark gray. In both sexes there is a slightly metallic iridescent sheen. Both sexes have a fringe on tibia I, which in the male extends on to the distal end of the patella as well. Length of female 4 to 6 mm; of male 4 to 5.5 mm.

This species is somewhat like those in *Tutelina,* but differs from them in that the height of the carapace is only half its width and only two-fifths its length; the carapace is slightly wider, the ocular quadrangle occupies half the length of the carapace; the sternum is relatively wider with its anterior truncature as wide as the base of the labium; the legs are not lineate with black; and the endites of the male are rounded apically, not drawn out to a point or hook. (Compare with Figure 639.)

New England and adjacent Canada west to Nebraska and Kansas.

27b Ocular quadrangle occupying only two-fifths the length of carapace **28**

28a Body and legs covered with dense pubescence. Carapace at least seven-tenths as wide as long and its height more than half the width . . . . *Metaphidippus* (35 species)

GENUS METAPHIDIPPUS

In general appearance the members of this genus resemble *Eris,* but the body is not quite as hairy, nor the cephalothorax quite as curved out at the sides. In many of our species the females have white scales diffusely spread over the carapace, while in the males, which are more strikingly marked, these scales are concentrated in a wide band on each side. There is much variation in pattern and degree of pigmentation making them difficult to separate by the inexperienced student. These are among the commonest jumping spiders collected by sweeping tall grass and bushes.

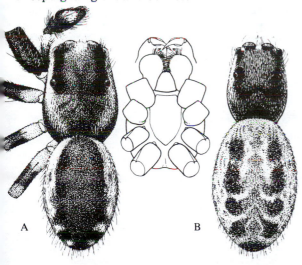

Figure 667. *Metaphidippus protervus.* A, Male; B, Female; C, Cephalothorax from below.

Figure 667. *Metaphidippus protervus* (Walckenaer)

There are four pairs of black to reddish spots on the abdomen. In some individuals the spots are joined to a central longitudinal band. Only occasional males show the four pairs of dark spots, but most have the abdomen dark brown, ringed with white. The cymbium and femur I have a dense covering of white scales. Length of female 3.7 to 6.3 mm; of male 3 to 4.4 mm.

Eastern States and adjacent Canada to the Rockies but more common in the north than south.

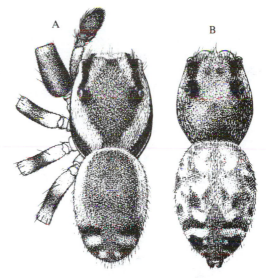

Figure 668. *Metaphidippus galathea.* A, Male; B, Female.

Figure 668. *Metaphidippus galathea* (Walckenaer)

On the abdomen of the female there is usually an indistinct chevron pattern or pairs of dark spots each preceded by a large white area. The fourth pair of white spots is transverse. The legs are conspicuously ringed. Some individuals resemble *protervus,* with which this has been confused. The cymbium and femur I of the male lack white scales or have only very few. Length of female 3.6 to 5.4; of male 2.7 to 4.4 mm.

Eastern States and adjacent Canada to the Rockies but more common in the south than the north.

Figure 669. *Metaphidippus aeneolus.*

Figure 669. *Metaphidippus aeneolus* Curtis

Similar in general appearance to the two preceding species. Length of female 5.5 mm; of male 5 mm.

Utah and Arizona west to the Pacific Coast States.

Figure 670. *Metaphidippus insignis,* abdomen.

Figure 670. *Metaphidippus insignis* (Banks)

There are five pairs of black spots on the abdomen, with the first pair sometimes indistinct. The abdomen is more yellowish than in the other species, and the legs are very faintly if at all ringed. The cymbium and femur I in the male lack white scales or have only very few of them. Length of female 3.6 to 5.7 mm; of male 2.9 to 4.7 mm.

New England and adjacent Canada west to Minnesota.

Metaphidippus manni (G. & E. Peckham)

The carapace is dark brown with white bands on the sides. The abdominal dorsum is light brown with a white basal band and with pairs of large black spots somewhat resembling

insignis. Length of female 4.2 to 4.8 mm; of male 3.3 to 4 mm.

Texas and Oklahoma west to the Pacific Coast States and British Columbia.

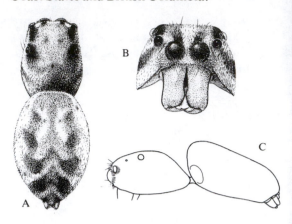

Figure 671. *Metaphidippus flavipedes,* A, female; B, face of male; C, lateral aspect of male to compare with *flaviceps,* Fig. 673D.

Figure 671. *Metaphidippus flavipedes* (G. & E. Peckham)

Except for the last pair, which may be brown, the legs are yellow and provided with a black band on the prolateral surface of femora, patellae, and tibae, most noticeable on I and II. Length of female 4 to 5.4 mm; of male 3.3 to 4.4 mm.

New England west to the Rockies. In Canada from Ontario west to Manitoba.

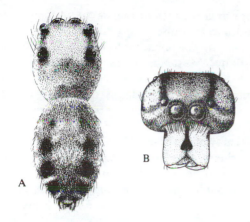

Figure 672. *Metaphidippus exiguus,* A, female; B, face of male.

Figure 672. *Metaphidippus exiguus* (Banks)

This species closely resembles *flavipedes* but the legs are without black lines and the abdomen has the pattern less distinct. Moreover, the chelicerae are yellow with a characteristic black mark as figured. Length of female 4 to 5.6 mm; of male 3.3 to 5.1 mm.

New England and adjacent Canada south to Georgia and Alabama west to Ohio.

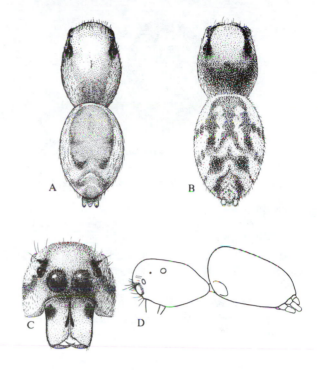

Figure 673. *Metaphidippus flaviceps,* A, male; B, female; C, face of male; D, lateral aspect of male to show elevated head region to compare with *flavipedes,* Fig. 671C.

Figure 673. *Metaphidippus flaviceps* Kaston

This species is generally similar to *flavipedes* but has the head region yellowish, quite shiny and very much elevated, especially in the male. This bulbous head region enables immediate recognition and separation from *flavipedes* and *exiguus.* Length of female 4 to 5.7 mm; of male 3.6 to 4.5 mm.

New England and New York.

Figure 674. *Metaphidippus canadensis*, abdomen.

Figure 674. *Metaphidippus canadensis* (Banks)

This species is darker than the others with only very narrow light chevrons on the abdomen. On the venter are three black longitudinal lines, the middle one sometimes not as distinct as the other two. Length of female 4.6 to 6.4 mm; of male 3.9 to 5 mm.

New England and adjacent Canada south to the mountains of North Carolina and Tennessee, west to Iowa and Minnesota.

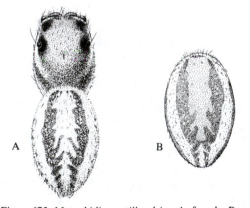

Figure 675. *Metaphidippus tillandsiae*, A, female; B, male.

Figure 675. *Metaphidippus tillandsiae* Kaston

The carapace is orange brown, darker along the sides and posterior declivity. On each side is a curved band of white scales extending from the anterior lateral eyes. The abdominal dorsum has a broad white band on either side continued as a thin basal stripe around the front. There is a dark brown band on each side having an irregular inner edge with an indication of chevrons behind. Between these is a median

orange area. Length of female 3.6 to 5 mm; of male 3.4 to 4.3 mm.

South Carolina to Florida and west to Louisiana. Most specimens have been taken from Spanish moss, and apparently the distribution of the spider is closely linked to those areas in which the moss will grow.

28b **Body and legs not densely pubescent; with iridescent scales. Cephalothorax less than two-thirds as wide as long, and its height about half the maximum width ....**
.... *Euophrys* (4 species)

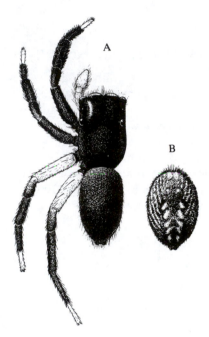

Figure 676. *Euophrys monadnock*. A, Male; B, Abdomen of female.

Figure 676. *Euophrys monadnock* Emerton

The female has the eye region black, the remainder of the carapace brown, the legs tan and unmarked. On the abdomen is a pattern of tan paired spots on a gray mottled dorsum. The male has the abdomen all black and the legs with contrasting black and orange-yellow as figured. Length of female 4 to 5 mm; of male 3.6 to 4 mm.

New England and adjacent Canada, the Rocky Mountain and Pacific Coast States.

FAMILY LYSSOMANIDAE

This is a small family with a single genus and one species in our area.

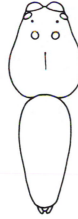

Figure 677. *Lyssomanes viridis.*

Figure 677. *Lyssomanes viridis* (Walckenaer)

The general color is a pale green, with black around the second, third, and fourth rows of eyes and some red scales in the head region. The chelicerae of the female are vertical, but those of the male extend out almost horizontally from the clypeus and are as long as the carapace. Length of female 7 to 8 mm; of male 5 to 6 mm.

This species may be found in low bushes.

North Carolina south to Florida and west to Texas.

Index
and Pictured Glossary

A

ABDOMEN: the second, or posterior, of the two major divisions into which the body of the spider is divided.
Acacesia hamata, 144
Acanthepeira stellata, 142
Acarina, 33
ACCESSORY CLAWS: the serrated bristles near the true claws on the tarsi of some spiders, 19
Achaearanea globosus, 106
 porteri, 106
 rupicola, 106
 tepidariorum, 105
Actinoxia versicolor, 71
Agassa cerulea, 249
 cyanea, 249
Agelenidae, 49, 56, 164
Agelenopsis aperta, 170
 naevia, 169
 pennsylvanica, 168
Agroeca emertoni, 219
 ornata, 219
 pratensis, 220
 trivittata, 219
Aliatypus californicus, 64
 thompsoni, 64
Allepeira conferta, 147
Allocosa funerea, 189
Alpaida calix, 153
Amaurobiidae, 44, 83
Amaurobius borealis, 87
 ferox, 87
Amblypygida, 35
Anahita animosa, 225
Anal tubercle, 21
Anatomy, 14
Anelosimus studiosus, 109
 textrix, 109
ANNULATE: as of a leg, showing rings of pigmentation.
ANTERIOR: toward the front
Antlike spiders, 8
Antrodiaetidae, 40, 61
Antrodiaetus pacificus, 62
 unicolor, 62
Anyphaena aperta, 224
 celer, 223

fraterna, 224
 maculata, 223
 pacifica, 224
 pectorosa, 224
Anyphaenidae, 51, 221
Aphonopelma baileyi, 67
 chalcodes, 67
 eutylenum, 67
 reversum, 67
Apollophanes, texanus, 239
APOPHYSIS: a process, heavier than a spine, usually applied to those on the legs or pedipalps, 17, Fig. 678.

Figure 678.

APPENDAGES: structures extending away from the body proper, as legs, palps, etc.
Aptostichus stanfordianus, 72
Arachnida, 14
Arachnidism, 25
Araneae, 29
Araneidae, 58, 130
Araneinae, 136, 137
ARANEOLOGIST: a biologist who specializes in the study of spiders.
Araneus andrewsi, 156
 cavaticus, 157
 cingulatus, 160

diadematus, 156
 gemma, 156
 gemmoides, 157
 guttulatus, 160
 juniperi, 158
 marmoreus, 158
 miniatus, 158
 niveus, 160
 nordmanni, 157
 partitus, 158
 pratensis, 160
 saevus, 157
 solitarius, 157
 trifolium, 157
Araniella displicata, 154
Arctosa littoralis, 189
 rubicunda, 189
Argiope argentata, 139
 aurantia, 138
 trifasciata, 139
Argiopidae, 130
Argiopinae, 137
Ariadne bicolor, 94
Arundagnatha, 162, 163
Asagena, 112
Attachment disk, 1
Attachment zone, 131
Atypidae, 40, 65
Atypoides gertschi, 63
 riversi, 63
Atypus bicolor, 66
 niger, 65
AUTOTOMY: the process whereby a spider will break off a leg being held by an enemy, 4
Aysha gracilis, 222
 incursa, 222
 velox, 222

B

Ballooning, 8
Ballus youngii, 249
Banded Garden Spider, 139
Basilica spider, 147
BASITARSUS: same as metatarsus, 18
Bathyphantes pallida, 126
Blabomma, 164
Black and Yellow Garden Spider, 138

Black Widow Spider, 99
 bite of, 25
BOSS: a smooth prominence generally at the lateral angle of the base of the chelicera, in some spiders, 15, Fig. 679.

Figure 679.

Bothriocyrtum californicum, 70
Bowl and Doily Spider, 118
Bridge-line of Web, 131
Bridge Spider, 155
BRISTLE: a long thin extension of the cuticle, more slender than a spine, 18, Fig. 680.

Figure 680.

Brown Recluse Spider, 89
Brown Widow Spider, 99
Burrowing Wolf Spider, 186

C

CALAMISTRUM: a series of curved bristles on the dorsal surface or retrolateral edge of metatarsus IV, of

some spiders, (the cribellate spiders), 19
Calilena restricta, 170
Calisoga longitarsus, 68
Callilepis eremella, 202
  imbecilla, 202
  pluto, 202
Callioplus euoplus, 87
  tibialis, 87
Callobius bennetti, 85
  nevadensis, 86
  nomeus, 86
  pictus, 86
  severus, 86
Calymmaria californica, 173
  cavicola, 173
  emertoni, 173
Camouflage, 8
Cannibalism, 5
Caponiidae, 47
Carapace, 15
Cardiac Area, 16, 20
CARINA: a keel, as on the clypeus (Fig. 681), or the chelicerae, in some spiders.

Figure 681.

Castianeira amoena, 218
  cingulata, 217
  descripta, 217
  gertschi, 218
  longipalpus, 217
  occidens, 218
Cave Orb Weaver, 135
Cellar Spider, Long-bodied, 96
Cellar Spider, Short-bodied, 95
CEPHALOTHORAX: the anterior of the two major divisions into which the body of a spider is divided, 15
Cercidia prominens, 142
CERVICAL GROOVE: the furrow which extends forward and toward the sides, from the center of the carapace and marking the boundary between the head and the thorax. It is sometimes indistinct or completely lacking, 15
Cesonia bilineata, 204
  classica, 204
CHELA: a pincer-like arrangement, as of the fang with the lamella on the basal segment of the chelicera, 45
CHELICERA: the front jaws, consisting of a strout basal segment, and a terminal fang, 15
Chiracanthium inclusum, 214
  mildei, 214
Cicurina arcuata, 167
  brevis, 167
  robusta, 167
  utahana, 167

Clasping Spines, 22
CLAW: a strong curved process at the distal end of the leg. There are at least two, and these are of about equal size. If a third is present it is median and ventral to the paired ones, and smaller and difficult to see. In some females the palp also has a claw, 19
CLAW TUFTS: the bunch of hairs at the tip of the tarsus in those spiders with only two claws, 19, Fig. 682.

Figure 682.

Clubiona abbotii, 215
  kastoni, 215
  maritima, 215
  moesta, 215
  obesa, 215
  riparia, 214
  tibialis, 215
Clubionidae, 53, 210
Clubionoides excepta, 216
CLYPEUS: the space between the anterior row of eyes and the anterior edge of the carapace, 15, Fig. 683.

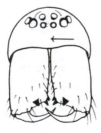

Figure 683.

Cocoon, 1
Coelotes juvenilis, 172
  montanus, 172
Collecting, 9
Color, Change of, 8
COLULUS: an appendage, usually slender and pointed, somewhat simulating a diminutive spinneret, lying between and in front of the anterior spinnerets, in some spiders, 21
Comb, 19
Comb-footed spiders, 97
Commensalism, 5
Conopistha trigona, 103
  rufa, 103
Coras lamellosus, 172
  medicinalis, 172

Coriarachne brunneipes, 233
  floridana, 233
  lenta, 232
  utahensis, 232
  versicolor, 232
Corythalia aurata, 253
Courtship, 5
COXA: the segment of the leg (or pedipalp) nearest the body, 17
Crab spiders, 227
Cribellatae, 42, 72
CRIBELLATE: adjective referring to a spider that possesses a cribellum.
CRIBELLUM: a spinning organ placed as a transverse plate just in front of the spinnerets, in some spiders. From it issues the so-called hackled band threads, 21, 42
Cross spider, 156
Crustulina altera, 110
  sticta, 110
Cryphoeca montana, 166
Ctenidae, 51, 225
Ctenium banksi, 111
  riparius, 111
Ctenizidae, 41, 68
Ctenus hibernalis, 226
CUCULLUS: a hood fitting over the anterior end of the animal, as in the Ricinulida.
Cybaeus reticulatus, 165
Cyclocosmia torreya, 69
  truncata, 69
Cyclosa conica, 147
CYMBIUM: the tarsus of the male pedipalp hollowed out to contain the copulatory organ, 17, Fig. 684.

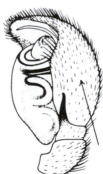

Figure 684.

D

DECLIVITY: an incline, as that which occurs at the rear of the carapace.
DENTICLE: a smooth tooth of small size usually on chelicerae, legs or palps, Fig. 685.

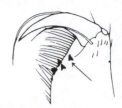

Figure 685.

DIAD: a pair, as of two eyes placed close together.
Dichotomous key, 27
Dictyna annulipes, 82
  bellans, 82
  calcarata, 83
  coloradensis, 81
  foliacea, 82
  hentzi, 82
  reticulata, 82
  sublata, 81
  volucripes, 81
Dictynidae, 45, 78
Digestion, 5
Diguetia albolineata, 90
  canities, 90
Diguetidae, 46, 90
Dinopidae, 43
Dionycha, 30
Dipluridae, 41, 67
Dipoena nigra, 110
Dolomedes scriptus, 179
  sexpunctatus, 179
  tenebrosus, 179
  triton, 179
  vittatus, 180
Domestic spider, 105
Dorsal furrow, 15
Dorsal groove, 15
DORSUM: the back or upper surface.
Drag line, 1
Drapetisca alteranda, 117
Drassodes gosiutus, 203
  neglectus, 202
  saccatus, 203
Drassyllus agilis, 206
  aprilinus, 206
  depressus, 205
  fallens, 205
  frigidus, 206
  insularis, 205
  niger, 205
  rufulus, 206
  virginianus, 205
Dugesiella hentzi, 67
Dysdera crocata, 93
Dysderidae, 48, 93

E

Ebo californicus, 236
  latithorax, 235
  mexicanus, 236
  parabolis, 236
  pepinensis, 235
Ecribellatae, 42, 87
ECRIBELLATE: adjective applying to a spider not provided with a cribellum.

Eggs, number of, 4
Egg sac, 2, 3
Eidmanella pallida, 115
EMBOLUS: the portion of the
  male copulatory organ
  through which the sperms
  are passed into the seminal
  receptacle of the female,
  Fig. 686.

Figure 688.

Erigonidae, 128
Eris aurantia, 259
  marginata, 259
Eriophora edax, 143
Ero leonina, 176
Estrandia grandaeva, 124
Euagrus comstocki, 68
  ritaensis, 68
Euophrys monadnock, 264
Euryopis funebris, 103
  limbata, 103
Eustala anastera, 143
  rosae, 143
Evarcha hoyi, 242
Eyes, 15

Figure 686.

ENDITE: one of the mouthparts,
  ventral to the mouth opening and lateral to the lip, so
  that in chewing it opposes
  the chelicerae, 17, Fig. 687.

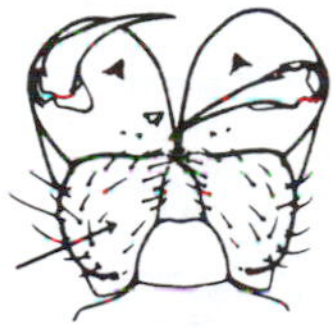

Figure 687.

Enemies, 13
Enoplognatha lineatum, 112
  marmorata, 112
  ovata, 112
  redimitum, 112
Entelogynae, 30
Envenomation, 25
EPIGASTRIC FURROW: a groove
  separating the region of the
  book lungs in labidognath
  spiders from the more
  posterior portion of the
  venter, 20
EPIGYNUM: a sclerite
  associated with the
  reproductive openings in
  the female of most
  entelogyne spiders. It lies
  in the midline just in front
  of the epigastric furrow, 22,
  Fig. 688.

F

Families, list of, 29
Family, key to, 37
FANG: the claw-like distal seg-
  ment of the chelicera, 15,
  Fig. 689.

Figure 689.

Feather-legged spider, 76
FEMUR: the third segment of
  the pedipalp or leg, count-
  ing from the proximal end
  of these appendages, 18
Filistata arizonica, 74
  hibernalis, 74
  utahensis, 74
Filistatidae, 43, 73
Filmy dome spider, 122
Fishing Spiders, 178
Florinda coccinea, 124
Flower Spider, 230
Flying Spiders, 1
FOLIUM: a pigmented design or
  pattern on the abdominal
  dorsum, 20

Foundation Line, 130
Foundation Zone, 130
Food of spiders, 4
Free Zone, 130
Frontinella communis, 118
  pyramitela, 118
Funnel Web, 1, Fig. 424
Funnel Web Tarantulas, 67
Funnel Web Weavers, 164
Furrow spider, 154

G

Garden Spiders, 156
Gasteracantha cancriformis, 132
  ellipsoides, 132
Gasteracanthinae, 132
Gea heptagon, 137
GENICULATE: bent in a right
  angle, as in the case of the
  base of the chelicerae in
  some spiders.
Genus, 26
Geolycosa missouriensis, 187
  pikei, 187
  turricola, 187
Gnaphosa brumalis, 201
  californicus, 201
  muscorum, 201
  parvula, 201
  sericata, 201
Gnaphosidae, 53, 200
Goldenrod Spider, 230
Golden-silk Spider, 136
Grass Spider, 167
Gray Cross Spider, 155

H

Habitats, 7
Habrocestum pulex, 241
Habronattus agilis, 243
  borealis, 243
  coronatus, 244
  decorus, 244
  viridipes, 242
Hahnia cinerea, 174
Hahniidae, 53, 174
Hamataliwa grisea, 199
Hammock Spider, 121
Haplodrassus bicornis, 209
  chamberlini, 209
  hiemalis, 209
  signifer, 209
Haplogynae, 30
Helophora insignis, 125
Hentzia ambigua, 251
  palmarum, 251
Herpyllus blackwalli, 205
  ecclesiasticus, 204
  propinquus, 204
Hersiliidae, 54
Heterogeneous eyes, 15
Heteropoda venatoria, 226
Hexura picea, 65
Hibernation, 5
Hololena curta, 171
  hola, 171
Homalonychidae, 50, 209
Homalonychus theologus, 210
Homogeneous eyes, 15
House Spider, 105
Hub of Web, 130
Hypochilidae, 42, 73

Hypochiloidea, 29
Hypochilus bonneti, 73
  gertschi, 73
  petrunkevitchi, 73
  thorelli, 73
Hypselistes florens, 128
Hypsosinga pygmaea, 152
  rubens, 152
  variabilis, 152
Hyptiotes cavatus, 77
  gertschi, 78

I

Icius hartii, 260
Ingestion, 5
Irregular mesh, 1
Ixeuticus martius, 85

J

Jumping Spiders, 240

K

Keys, use of, 27
Keys to Families, 36
Kibramoa guapa, 91
  suprenans, 91

L

Labelling, 12
Labidognatha, 38, 72
LABIUM: the lower lip, between
  the two endites, 17,
  Fig. 690.

Figure 690.

LAMELLA: a triangular plate on
  the promargin of the
  cheliceral fang furrow in
  some spiders. It resembles
  a broad tooth, and forms a
  kind of chela with the fang,
  45
LAMELLIFORM: flattened, as of
  certain hairs in claw-tufts,
  Fig. 566.
LAMINA: a flat plate, Fig. 691.

Figure 691.

Index and Pictured Glossary   269

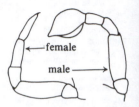

Figure 695.

Figure 696.

R

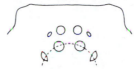

Figure 697.

S

Figure 698.

Figure 699.

Figure 700.